Edward Jesse

Lectures on Natural History

Edward Jesse

Lectures on Natural History

ISBN/EAN: 9783337025830

Printed in Europe, USA, Canada, Australia, Japan

Cover: Foto ©berggeist007 / pixelio.de

More available books at **www.hansebooks.com**

LECTURES

ON

NATURAL HISTORY.

BY

EDWARD JESSE, Esq.

AUTHOR OF "GLEANINGS IN NATURAL HISTORY," "ANECDOTES
OF DOGS," "FAVOURITE HAUNTS AND RURAL STUDIES,"
"AN ANGLER'S RAMBLES," "WINDSOR AND ETON,"
ETC. ETC.

DELIVERED AT

THE "FISHERMAN'S HOME," BRIGHTON.

LONDON:
L. BOOTH, 307 REGENT STREET, W.
1861.

LONDON:
STRANGEWAYS AND WALDEN, PRINTERS,
28 Castle St. Leicester Sq.

These Lectures

ARE DEDICATED TO

THE BRIGHTON FISHERMEN,

BY THEIR

SINCERE FRIEND AND WELL-WISHER,

EDWARD JESSE.

INTRODUCTION.

Some of the readers of the following Lectures might wish to be informed what induced the Author to write and deliver them to the Brighton fishermen. It may, therefore, be stated, that the occupation of these men kept them very frequently for days together at sea, often on Sundays. The consequence was, that they became much neglected, both as to their temporal as well as their spiritual improvement. Indeed, there was a time when a part of the beach at Brighton exhibited a scene of quarrelling, swearing, and drunkenness, such as was seldom witnessed. It is far different now, for, owing to the exertions of some benevolent persons, an arch, capable of containing about eighty persons, was hired for their use. This has been floored, white-washed, lighted, and warmed; and the walls covered with many amusing and instructive drawings and coloured prints, presenting a very cheerful aspect. Seats and small tables have also been provided, together with a library of amusing and instructive books, in addition to newspapers and periodicals.

When the "Home" was thus completed, the Author was requested to inaugurate it with a Lecture, which he willingly undertook to do. It was listened to by a most attentive audience, and suitable addresses were also made to the men by his friend Mr. Cordy Burrows, and other gentlemen present. These Lectures were continued from time to time, as occasions offered, during the period the Author remained at Brighton, and last winter he resumed the practice; and he has reason to be gratified with the result of his labours, for his assertion will be borne out by many others, that a more well-conducted, sober, and attentive set of men than his hearers have seldom been collected together.

But his success must not be attributed exclusively to the Lectures. The fishermen have been induced to abandon the ale-houses and beer-shops (those curses of labouring men), in consequence of being amply and liberally supplied with good hot coffee at all times of the day, and until nine o'clock at night. This, they find, does them more good than either spirits or ale, and they thus avoid those miseries and that poverty which drunkenness is sure to bring, with all its curses, on themselves and their families. Thus, by the judicious use of coffee, the men have become a sober class, and are enabled to make deposits in Savings' Banks, and can therefore feel that they are independent of sickness and occasional want of employment. This fact is one of great social importance, and might be beneficially followed in every town and village in England. Indeed, those benevolent persons, especially females, who visit the dwellings of the poor, might teach them how to roast, grind, and make coffee. All that is wanting is an iron tray, and a wooden pestle and mortar, for the purpose, and they would cost but a mere trifle. These, with a pound or two of coffee-berries, and a lesson for the use of them, would be an acceptable and useful wedding-present for the bride of a working-man, and might tend to win her husband from resorting to an ale-house.

But to return to the Lectures. They were written partly from the Author's own notes; and were, in part, extracted from different works on natural history. Some of them have been read and approved of by the Author's friend and neighbour, Professor Owen; and he, therefore, submits them with some little confidence to the public. He may add, that the instruction contained in them may be found useful in schools, and to the young generally. Where they have been read to children they have been listened to with attention and pleasure.

EDWARD JESSE.

East Sheen, Surrey,
July 1861.

[From the *Brighton Herald* of January, 1860.]

At the foot of the Ship-street "Gap," Brighton, there is a series of arches built in the face of the cliff; and in one of these a very novel and interesting scene took place on Thursday evening. It is known as the Fisherman's Home and Reading-room, and is a warm, comfortable, well-lighted place, perfectly adapted to induce the beach population to prefer a cup of coffee and a book to the expensive luxury of the beer-shop. In this room were gathered as many seamen as could be crowded into it, this being a kind of inaugural assembly,—one of, we hope, a long series of nights on which the members of the Home will meet for readings, songs, and other means of rational amusement.

As this was the first, so it was a special night, marked by the attraction of a special paper written for the occasion by Mr. Jesse. He did not, however, read his own paper.

Alderman Burrows undertook that task; but, before commencing the reading, he took occasion to observe that this Institution had been provided by persons taking an interest in those who were not so well able to help themselves, in order that the fishermen frequenting this beach might have a nice, warm, dry, comfortable apartment, in which they might read newspapers and books provided for them, so that they might be led to make an intelligent and right use of those faculties which the Almighty had bestowed upon them. By educating himself, a man was the better able to appreciate the advantages of a good and virtuous life, and also to discriminate between right and wrong, so that he might do his duty towards his neighbour, his duty towards himself; and in the performance of those two duties he would be doing the greatest of all duties, that towards his Maker. It was with this conviction that this Reading Room had been provided; and it gave him much pleasure to see it so full this evening. If the fishermen only showed their appreciation of it by their attendance, an adjacent arch would also be fitted up, and there would thus be accommodation for double the number. In the meantime, as many among the fishermen had not the advantage of education, it had been thought desirable that there should be some one to read aloud some amusing book in the room of an evening: he had pledged himself to do so once a-fortnight, and if other gentlemen would do the same, there would, with very little trouble, be an amusing and instructive entertainment provided there almost every evening in the week. It was a source of pleasure to find many persons taking a warm interest in this Institution, which was to a considerable extent self-supporting,—and it is always a pleasure to help those who help themselves. He was accompanied by a very dear and affectionate old friend, Mr. Jesse, who had kindly prepared a paper on "Singular Facts Relating to Fish," *expressly for this occasion.*

[From the *Brighton Herald* of February 6th, 1860.]

"Really our fishing population have reason to be grateful—they ought to consider themselves highly favoured—when a gentleman of Mr. Jesse's talents and years, visiting Brighton for the benefit of his health, takes so much interest in their behalf as to write for their especial use and instruction such valuable and interesting lectures as he has done; and, moreover, to have those lectures printed and sold, as he promises to do, for the benefit of their 'Home.' The men evidently *do* appreciate Mr. Jesse's kindness, as a token of which they have returned their thanks to him as follows:—

"'We, the undersigned fishermen and boatmen of Brighton (and others present), beg to return our sincere thanks to Mr. Jesse for writing and reading his good lectures.

(Signed)

"'Joseph Salvage, Richard Markwick, George Young, Jim Shrivell, William Pentecost, Thomas Pentecost, George Priest, and Philip Collins (fishermen).

Joseph Wells, James Harries, and Friend Payne (seamen).

Thomas Laycock, John Measor, Francis Measor, Richard Gillam, Thomas Daws, James Mockford, William Bray, Samuel Akehurst, George Gunn, John Gooding, George Harman, Philip Barnard, Thos. Bassett, John Laycock, James Bassett, Thomas Bassett, sen., George Monk, John Barton, and George Jeffery (boatmen).'"

LECTURES, &c.

I.

SINGULAR FACTS RELATING TO FISH.

My dear Friends,—

I am going to read you something about fish, as you are all fishermen; but not all about the fish of this country, as you know them well, but about the fish of hot countries; and what I am about to say of them will, I am sure, interest you. The facts I am going to state are well known to observant naturalists, and are sufficiently authenticated to leave no doubt of their truth. In the East Indies, Ceylon, and other hot countries, there are numerous ponds and tanks, which in the rainy season are well filled with water as well as with fish. When the sun afterwards blazes forth in all its tropical heat, these ponds and tanks are dried up, and you would suppose that the fish would also be dried up with them. But this is not the case. The fish have the power of penetrating deeper and deeper into the mud as it gradually dries, till they find a sufficient moisture to keep them alive till the periodical rains come some months afterwards, when they find their way back again into the water. In the meantime, their retreat is occasionally disturbed by the natives, who dig down and find them generally at a depth of two feet. The soil is clay, into which they have the power of burying themselves. In some of the sandy plains of the East Indies there are large but shallow ponds, or rather hollows, which are filled in the rainy season, but in hot weather are perfectly dry. As they become so, great numbers of small fish may be seen dead on the sandy surface; but on the recurrence of heavy rains these hollows are again stocked with fish.

Now, Dr. Buist, a learned and observant naturalist, gives us many instances of fish having fallen from the clouds. He tells us that, on the 19th February, 1830, at noon, a heavy fall of fish occurred at Nokuthatty factory, and that attestations of the fact were obtained from nine different parties. The fish were all dead; some fresh, others not. They were seen in the sky like a flock of birds, descending rapidly to the ground. Again, he says that, on the 16th and 17th May, 1833, a fall of fish occurred near Futtehpoor, after a violent storm of rain. The fish were from 1½lb. to 3lbs. in weight, and all dead. Some fish fell at Meerut, while Her Majesty's 14th Regiment were out at drill, and were caught in numbers. At another time, during rain, a quantity of live fish, about 3 in. in length, fell about 20 miles from Calcutta; they were all of the same kind. Many other instances might be mentioned; one vouched for by the late Governor of Ceylon, Sir Emerson Tennent, who, while riding out after a heavy rain, observed many small fish alive on the road. After these facts, I should mention that, if showers of fish are to be explained, it must be on the assumption that they are carried up by squalls or violent winds from rivers or spaces of water not far away from where they fall. An instance also occurs to me, in this country, which was seen by a friend of mine. He had a garden surrounded by a high wall, with a dry soil. After a heavy fall of rain, the garden was filled by myriads of small frogs, which must have descended from the clouds.

After these remarks on the showers of fish, I may tell you what is still more extraordinary — that some fish in Ceylon, in the dry season, leave the pools when they contain but little water, and can make their way through the grass to other pools, going to them in one direct line. Some fish, also in Guiana, have been seen travelling overland during the dry season, in search of their natural element, in such droves that the negroes have filled their baskets with them. Sir John Bowring, in his account of the embassy of the Siamese kings, in 1855, states that in ascending and descending the river Meinam he was amused with the curious sight of fish leaving the river, gliding over the wet grass, and losing themselves amongst the trees of the jungle. Whilst travelling on land, the fish have their gills open or expanded. The class of fish which have this power of moving on land have some of their bones so disposed in plates and cells as to retain a supply of moisture, which, while crawling along, gradually exudes so as to keep the gills damp. Another small species of fish is often seen travelling along a hot and dusty road in Ceylon, under the mid-day sun, in search of water. Extraordinary as these facts may appear to you, they are perfectly well vouched for; also, that a species of perch in Ceylon, of a very peculiar formation, has been seen to ascend trees, in search probably of some food,—insects most likely.

Now, in hearing what I have to say to you, I must request you to bear in mind that a Benevolent Creator has endowed animals, fish included, with that formation and those instincts which are necessary for their self-preservation. I will give you a proof of this which is familiar to you. You know that many flat fish,—soles, turbot, &c.,—have brown and white surfaces. When they are attacked by other fish which prey on them, they remain flat on the white surface of their skins, showing only the brown surface, which is generally the colour of the sands on which they feed; and thus they escape the observation of their enemies. So it will be found to be all through Nature. Every animal is furnished with either some means of escape, of defence, or of cunning. Some are swift, and some are strong, and others hide themselves from their enemies in holes in the earth. But all are fed in some way or other by the same Almighty hand which created them. You will find a beautiful reference to this in Psalm cxlv.:—

"The eyes of all wait upon Thee, O Lord; and Thou givest them their meat in due season.
"Thou openest Thine hand, and fillest all things living with plenteousness."

If God thus care for the birds of the air, the beasts of the field, and the fishes of the sea, be quite sure that He will take care of you if you place trust in His goodness. When out at sea in your boats, exposed perhaps to boisterous and contrary winds, as most of you must have been, then lift up your hearts to that Good Being who can make the storm to cease and the waves to be still. You will always find this a great comfort and relief, especially when you are in any danger.

Let me now give you an instance or two of the goodness of Almighty God to His creature man. You are aware that mackerel, herrings, pilchards, sprats, &c., are caught in vast numbers at certain seasons of the year, though perhaps you may not be aware that there are as many hills, and, probably, mountains, in the sea as there are upon dry land;—this has been partly ascertained in laying down the cable across the Atlantic. Now, these hills are covered with marine plants, like a forest or jungle; and you might suppose that, for purposes of concealment or protection, these gregarious fish would resort to such places to deposit their spawn; and if they did, you may well suppose how few would be taken in your nets. Now, a benevolent Providence has designed that these fish should become useful to man. And so they are, to a great extent; for, besides affording food to many in this country, they are exported to different parts of the world. I must now tell you that no spawn of fish will vivify, or become productive, without light. There is, then, a powerful instinct implanted in them, which compels them to resort to shallow places in the sea, in order to deposit their spawn under the influence of light, and where you are enabled to use your nets to advantage. These gregarious fish, or fish which go in shoals, are pursued by many enemies besides yourselves: gulls, porpoises, and other fish follow and feed on them, and yet nothing drives them from the shoals. It is the incredible numbers alone which keep up the annual migration.

Before I conclude, I wish to call your attention to an unusual occurrence which took place along our coasts about four years ago, and which some of you may, perhaps, be able to give some explanation of. At the time referred to a very heavy fall of snow took place, and soon afterwards great numbers of conger-eels and many other fish were picked up perfectly blind. Now, some of these eels and fish were examined by Professor Owen, the head of the Natural History Department in the British Museum, and second to no one as a scientific anatomist. His report was that the eyes of the fish he examined were covered with an opaque substance, which produced total blindness. This curious fact was not confined to fish off our Brighton coast. At Southampton, and many other places, grey mullets and other fish came on shore perfectly blind; and I picked up what, I believe, is called the parrot-fish, in the same state. It is not easy to account for this sudden blindness. I am aware that some attribute it to the fall of snow; but why conger-eel and other deep-sea fish should be affected by it is a mystery not easy to be solved. I shall, therefore, be glad to hear what any of you have to say on the subject. Let me add that fishermen enjoy the advantage over landsmen of seeing the sun rise on a beautiful morning, shedding its gilded rays over the rippling waves far out at sea, and also of viewing the glorious setting of the same luminary. These are sights which may well excite your admiration, and which ought to raise your thoughts to that Great and Good Being who has preserved your lives in the midst of many and great dangers, and at this moment enables you to meet together in peace to receive instruction and amusement. May He bless you all!

II.

SECOND PAPER ON FISH.

My dear Friends,—

In my last lecture to you I endeavoured to address you in that mood and way which would best come home to your feelings, and show you that I had only one object in view, that of amusing and improving you. In fact, I wished you to consider me as your true friend, and such, I hope, I shall be found to be. You appeared to listen attentively, as well as to like what I wrote for you, a few evenings ago, on the subject of fish in hot countries. Some of you might have thought that what I then said respecting fish making their way over land to other waters and of their climbing up trees was not very probable; nor can I wonder at it. Now, I wish you to consider for a moment the possibility of fishes being differently constructed or formed in some countries to what those are which you take in your nets off the Brighton coast; and in so doing always bear in your minds that Divine Wisdom knows no bounds. His will is the well-being and well-doing of all His creatures, each of them in its own place or sphere, and His Almighty power enables Him to give peculiar formation and faculties to those beings which His wisdom created and His will decreed.

After these remarks, I may tell you that it is by a muscular movement of their ribs that fish can propel themselves along dry land, somewhat in the same way that eels and serpents are known to do. Now, if you will look at this sketch of a fish I have brought with me, you will see that what may be called ribs can be used as organs of muscular motion, according to the will of the animal. I should be making this lecture of too great a length if I were to enter more minutely into the particulars of the anatomy of these crawling fishes; but I hope enough has been said to convince you that they exist. As to the fact of fish ascending trees, I will only mention one proof of it, which is, that there is a little climbing perch well known to naturalists, and which is found in the mangrove swamps. It can ascend trees like a chimney-sweep, and this it does by only using a pair of prickles from the gill-flaps, instead of elbows, and thus it gains the tops of stems many feet above high-water mark, picking off the flies that alight on the tree it climbs up.

But let us now turn to flying fish. There are several different species of these, and they are sadly persecuted, being pursued in the sea by bonitos and other rapacious fishes, and take flight when they are in danger from them. While in the air, the frigate and other marine birds make them their prey. It may seem to you surprising that there should be flying fish as well as creeping and climbing fish, and yet they are known to exist.

Many of you, no doubt, have heard of the sea-cow. It appears a mass of blubber and almost incapable of motion, yet its fins are covered with a sort of nails which enable it to crawl on shore, assisted by its pectoral fin, and also to get on the ice. I mention this to prove to you that the fins of these creatures are as curious as the formation of the fish referred to, and that in both cases the fins of the sea-cow, and the muscular movement of the bones of crawling or climbing fish, serve the purposes of feet.

I now wish to call your attention to some of the enormous monsters which are to be found in more distant seas; for, fortunately for you, you have not to encounter them in your nets off these shores. One of these is called a squill. It is provided with several arms of enormous length, and when living it is said to be as transparent as crystal. It has a large mouth, and its eyes are of a sky-blue colour, embedded in the substance of the head. This monster is said to form part of the food of the whale. I will now relate an anecdote of the squill. The captain of a whaler landed on a small, uninhabited, rocky island in the South Pacific, with one of his mates, in search of curious shells. The tide was receding, and the mate, having gone a few feet up a rock, found a squill adhering to it. Never having seen one of them before, he disturbed it, when the creature endeavoured to flounder down to the sea. The man intercepted it in its course, when it raised itself up, and seized him with its long arms, squeezing him in such a way that he felt as if all his bones would be broken, at the same time it breathed hot air into his face, and glared at him with its blue and angry eyes. In this extremity the mate called out to the captain, who luckily was near, and who came and released him by cutting in two the arms of the squill with a large knife which he had with him. Had it not been for this interference, the man would have been killed and his body fed upon afterwards. In the Mediterranean these creatures are found, but of an inferior size. They spread themselves on the ground, and, when persons are bathing, instances have been known of their being seized and killed by the squills.

There is another monster found in the West Indies called the sea-eagle, because in its rage and anger it sometimes elevates itself with such force as to raise the sea into a foam, and makes a noise like thunder. One of the species, called by sailors the sea-devil, was taken at Barbadoes, and was so large that it required seven pairs of oxen to draw it on shore. Sharks and rays, which are nearly allied to them, are known to have been caught of the enormous length of forty feet.

You may well ask, What can be the use of these and many other monsters of the deep? I will endeavour to explain this to you. You know that quadrupeds in general only have one young at a time, while a fish will produce a million; and, indeed, it is calculated that a codfish alone will put forth nine millions of eggs in one year. Now it is evident, that if it were not for these 'monsters, which open their enormous mouths and throats and swallow the smaller fish by hundreds, the ocean would soon be filled with them, and there would scarcely be room for other marine animals. Only think for one moment how gregarious fish, such as mackerel, herrings, &c., would increase and multiply if it were not for your nets, and the predatory creatures which feed upon them. Remember also that He, whose tender mercies are over all His works, has fitted the creatures thus exposed to destruction for their fate; and we may therefore conclude that, being what are called cold-blooded animals, they do not suffer from great pain and anguish. The tremendous animals I have referred to also devour all carcases, &c., which may be found floating in the water, and thus they serve to purify the ocean, as hyænas and vultures do upon earth. I will mention another lesson which may be learnt from the existence of these monsters; for, if God fitted them to devour, He fitted them also to instruct. The existence of creatures so evil, and such relentless destroyers of the works of the Almighty, teaches us that there are probably analogous beings in the spiritual world, and which should warn us to use great care and watchfulness in our conduct, in order that we may escape their destructive fury. You see, I occasionally give a little good advice while I am endeavouring to amuse you.

Nothing is more remarkable than the infinite variety and singularity of the figures and shapes of fishes. It has been thought that the ocean contains representatives of every animal that is to be found on the earth, or in the air—at all events, the forms of fishes are more singular and extraordinary than those of any other department of natural history. Amongst the animals of South America, one of the most curious and interesting is the gymnotus, or electric eel —so strong is their electric power, that it is said they can kill a fish by it (and on which they feed) at a distance of sixteen feet. Lacepède, the celebrated French naturalist, is my authority for this statement. They abound in the rivers and ponds of South America. When the Indians want to catch them, they assemble and drive the wild horses of the plains, by shouts and other means, into the river. The electric eels then attack them: now and then a horse will receive so severe a shock that he is killed. Others contrive to swim across the river, and then throw themselves down exhausted on the opposite bank. The eels are thus deprived of their electric power for a time, and are then speared by the Indians, who feed on them. There is another singular fish, which is able to bring its prey within its reach by discharging a different element than that of the electric eel, and that is water. It is a small fish, remarkable for its singular shape, the brilliancy of its colours, and the quickness of its movements. It may be called the fly-shooter, from its food being chiefly flies and other insects, and especially those that are found on aquatic plants. When it sees one of these on a leaf, it blows out a drop of water with some force, which knocks the fly off the leaf, and it then feeds on it. I will mention another fish, and an ugly one it is, which you are probably acquainted with, for it is found in all the European seas. They are sometimes called fishing-frogs, from their resemblance to that animal; but I believe you call them sea-devils. It is a large fish, and has been caught seven feet in length. Now, this fish has no defensive arms, nor strength in its limbs, or quickness in swimming; but it is a cunning fish, as I will prove to you. In order to procure its food it hides itself in the mud, covers itself over with sea-weed, or conceals itself among stones, and lets no part of it be seen but the end of some fringes of its body, which it moves and agitates in different directions, so as to make them appear like worms or other baits. Fishes, attracted by this apparent prey, come near this sea-devil, when he catches them in his enormous throat, which is furnished with almost innumerable teeth. There is another of this species which has only a single, what may be called, bait, just above his mouth. You see in this case of the sea-devil, that if it cannot pursue and overtake and seize its prey, it is enabled—as in the case of the electric eel and the fly-shooter—to do so in a way we should not expect; thus showing the beneficence, wisdom, and power of the Great Creator, and to which I am always glad to call your attention.

I will only mention the habits of one or two other fish, as they may interest you. The hussar fish of Demerara, and the black goby of the Mediterranean, each makes an artificial and prettily-made nest; the first of fresh-water plants, and the other of sea-weeds. They protect their spawn and defend their young fry, observing in this way all the instincts of birds that lay eggs. The little sticklebacks of our brooks in England do the same. The little frog fishes have side-bags full of water, and pectoral fins, like feet. They hop about for hours on the sands, left dry by the retreating tide, to prey upon the sand-hoppers, &c. The mud-fish of the river Gambia, in Africa, when the stream falls low, burrows and coils itself up in a deep mud-cellar, leaving a little hole for air, which its swim-bladder deals with like a lung. ' When the rainy season returns it comes out, and then breathes as fish do. It is a true amphibious creature, that is, it can exist both on land and in the water.

I have now done, and trust that I have interested you. From my advanced age, I may not be spared to come amongst you another year; but while I live I shall always be delighted to hear that you duly appreciate and avail yourselves of the advantages prepared for you in your Fisherman's Home.

III.

ON BIRDS.

My dear Friends,—

I have given you two lectures on foreign fish; and, as they appeared to amuse you, I will now read you one on birds, because, except gulls and a few marine birds, you are not very likely, from your occupation as fishermen, to know much of their habits and peculiar instincts. These are well worth your attention, and I hope that what I shall have to say on this subject may both instruct and amuse you. But, before I proceed further, I wish to say a word respecting the gulls which dip and flit and fly about these shores, in a way which every lover of nature, and every visitor to this town, must always admire. Now I ask of you, as a little return for the trouble I have taken in writing these lectures for you, to protect these interesting birds as far as you are able to do so. I always watch their flight with the greatest satisfaction: it is one of the sights which renders my visit to Brighton a pleasing one. Besides, they are useful birds; for, if you take a stroll on the Downs when land is being ploughed up, you will perhaps see gulls following the ploughman, and picking up the grubs of many of those insects which are so injurious to the farmer.

As there are but few trees and bushes in the more immediate neighbourhood of Brighton, I am afraid that you are not often gladdened with the song of the blackbird, thrush, and nightingale, though you have probably heard the cheerful notes of the lark, as he pours them forth, and approaches to heaven nearer than any other bird. The lark makes its nest always on the ground, and generally early in the spring, in grass fields. If you examine the claws of a lark,—which you may do at Mr. Sinnick's or any other poulterer's, where, I am sorry to say, too many of them are to be found,—you will find that these claws will readily take up one of the eggs of these birds. When, therefore, the mowers either approach to or mow over their nests, they will take up an egg in each foot, as I have seen them do, and convey them to some more secure place, returning quickly for the rest, till all are removed. Now, in viewing the structure of the foot of the lark, one cannot help admiring the goodness of a benevolent Creator, who has thus supplied one of His creatures with the means of rearing its young. Should the egg be only just hatched, the young will be removed in the same way.

I will now tell you that there are about forty different sorts of tender and, generally, what are called soft-billed birds,—that is, birds that feed on insects,—which arrive in this country from far-distant places every spring. You may not be aware that these little birds perform their long journeys in the night, led by an unerring instinct, which the Great Creator has implanted in them. This fact I have ascertained from some of the keepers of lighthouses, who have informed me that they have occasionally found these birds, early in the morning, killed by flying against the revolving light. They have also found woodcocks, snipes, and other birds, dead, showing that they also migrated in the night. As another proof of this curious fact, I may mention that, riding out early one morning in a meadow, a large flock of swallows dropped on the ground near me, and so much exhausted that they appeared incapable of moving, although I rode my horse amongst them. After resting some time they took flight, and dispersed in various directions. Swallows are supposed to migrate to this country from Africa, Italy, Spain, Greece, and other places. A captain of a ship assured me, that when he was at a long distance from land, numerous swallows settled on his rigging, as a resting-place. They are a useful bird, destroying myriads of flies; for they are on the wing, catching them, from the first light of day till late in the evening, It is a pretty sight to see them thus employed. They are sensible, clever birds, and I will give you one or two instances of this. You know that swallows make their nests of mud or clay. Now sparrows are apt to take possession of these nests and lay their eggs in them. When the hen-sparrow is sitting on them, a number of swallows will collect together, each with some clay in its mouth, and, in an instant, stop up the hole of entrance, thus leaving the sparrow to starve to death in the nest she had stolen. This fact I observed myself, and also the following:—A pair of sparrows had driven a pair of swallows from their nest, laid their eggs in it, sat on them, and hatched six young ones. When this took place, a number of swallows came and pecked down their former nest, and I saw the helpless young sparrows on the ground, where they soon perished. A third instance of this combined intelligence in birds was communicated to me by the late Sir Henry Willock, who was our Ambassador in Persia. There was a ruined tower opposite his window, at Teharan, on which those migratory birds, the storks, came year after year to make their nest. On one occasion a pair of peafowl forestalled them, and took possession of the tower and began to prepare a nest, driving the old storks away. After a short time a number of these latter assembled, attacked the peafowl, drove them away, and remained near the spot until the original storks were securely established on the tower. Now you must perceive that this faculty of communicating their wants and of exciting their congeners, or others of their own species, to assist in revenging their wrongs, is not only curious, but wonderful. How this is done must be left to conjecture, except that it is an impulse implanted in them by their Divine Creator.

Dogs that have been ill-treated by a larger one have been known to entice another to revenge their cause.

I will now tell you a little about the cuckoo—a bird, I am afraid, you seldom hear at Brighton; but it arrives in this country early in the spring, and its unvaried notes seem to proclaim fine and pleasant weather. It is a lazy bird; for, instead of sitting on and rearing its young, as all other birds do, it deposits its eggs, but only one, in the nests of other birds, selecting always those of insectivorous birds, that is, of birds which feed their young only on insects: these are generally robins, wag-tails, and hedge-sparrows. Now, the cuckoo is as large as a blackbird, and requires a great quantity of food, It is evident, therefore, that if the parent robin, &c., had to feed their own young as well as the voracious cuckoo, some of them would be starved. In order to prevent this, the latter is furnished with a hollow in his back, in which he contrives to get the newly-hatched robin or hedge-sparrow, and then throws them out of the nest one by one, remaining sole possessor of it. Having done this, he is readily fed and brought up, though it requires all the exertions of his foster-parents in order to supply his enormous appetite. There is another curious fact connected with the cuckoo. There was a small hole in the wall of my garden, in which a robin had built its nest. Now, it was quite impossible that a cuckoo could get into it to lay its egg, and yet I found a young cuckoo in it. She must, therefore have dropped her egg on the ground near the hole, and either taken it up in her mouth, or in her foot, and placed it in the nest.

Perhaps you are not aware that there is a great difference between rooks and crows, although it is very usual to call them all crows. The rook feeds on worms, slugs, &c., and is very useful to the farmer; while the crow is not only a great thief, but will kill, if he can, other birds, in order to make them his prey. A gentleman driving one day in his gig along a lane in Shropshire, saw a house-pigeon pursued by two carrion crows,—so they are called from eating carrion,—as they were probably hungry, and wanted the pigeon for food. The latter, becoming exhausted, fled for refuge into a tall, thorny hedge. One of the crows, however, stationed himself above the pigeon, and the other below it. They then got nearer and nearer to the poor bird, who, seeing its danger, left the hedge, but was immediately followed and seized by one of the crows. To the surprise of the gentleman who had watched the whole proceeding, he saw the crow rise up into the air, and at last fall down dead; one of those active little animals, called weasels, had fastened on him. This serves to illustrate the old proverb of the biter being bit. The pigeon was afterwards picked up alive, and taken home by the gentleman referred to. I will now tell you another anecdote of the crow. In the Island of Ceylon there is a very cunning and impudent one, not black, as ours are, but with a brown or bronzed back. In the court-yard of the house of the Governor of Ceylon a large dog was chained up, and was employed one day in picking the meat off a bone which had been given him. One of these crows alighted near him, and showed a wish to partake of the bone. This the dog would not allow, so the crow flew away and soon returned with a companion, who was placed near the tail of the dog, and the other took his station as near as he could venture to the coveted bone. The crow near the dog's tail then gave a strong pluck at it, when the dog turned quickly round to see who had taken this liberty with him. This was the opportunity wanted, for the first crow seized the bone and flew away with it, followed by his companion, and they no doubt enjoyed it together in some secure place. You see this is another proof of what I have said of combined intelligence and communication in animals. The above curious anecdote is related by the late Governor of Ceylon in his history of that island.

There is another cunning bird, which Shakspeare calls the maggot-pie, but we the magpie. In a tame state they are easily taught to talk, which they will sometimes do quite as distinctly as a parrot; in a wild state they now and then make a chattering noise, but do not often collect together as rooks do, although some years ago I counted more than twenty in one flock on the Brighton Downs. They make curious nests, using a large quantity of thorny sticks and brambles, and sometimes place them in odd situations, as I am about to tell you. You have heard of the celebrated Dr. Johnson, who, almost without assistance, and in the midst of poverty, completed that wonderful Dictionary,— a proof of his great learning and extensive reading,—which is called after him. In his more prosperous days he went a journey into Scotland, accompanied by his friend Mr. Boswell; and as he was a large, heavy man, he had a stout stick in proportion to his size. Now, it so happened that he lost this stick, and deplored his loss very much. His companion, in order to comfort him, said that it would be found again. "Never!" said Dr. Johnson: "consider, sir, the value of such a piece of timber in this country!" Thus you see that trees were not very abundant in those days in Scotland. This being the case, a pair of magpies, for want of a better place, resolved to make their nest in a gooseberry-bush in a garden. They brought great quantities of brambles, sticks, and gorse, or furze, and made it so large, and with so many twistings and turnings in it, that no arm (even if it were as long as my own,—and that is pretty long, as you may see,) could reach either the eggs or the young ones. It was considered a curiosity, and was suffered to remain unmolested; and there it may be, for all I know, at present, although the Scotch are a thrifty people, and might not like to lose their crop of gooseberries.

I do not know whether any of you have seen the fishing eagle. It is sometimes found in this country, and it feeds upon fish when it can get any. It is a noble bird, but not so fine a one as the golden eagle, for which it is sometimes mistaken. A lady told me that a flock of wild swans, perfectly white, flew past her drawing-room window in Ireland, pursued by two eagles. It must have been a fine sight. As the swan is a much larger bird than the eagle, it shows the boldness and power of the latter in attacking such a prey.

The heron is a bird you have probably seen, and a most patient one it is, standing, with its long legs, in the shallow water of some pond or stream, for hours together, waiting to catch an eel, or some other fish, or a frog or water-rat. It will also feed on snakes. In frosty weather they go to the marshes, as well as to the sea-shore, in search of food. They build on the highest trees; and, notwithstanding the great length of their wings, they quit the trees and alight on them again with the most perfect silence. In the fine heronry in Windsor Great Park—and it is a royal one—I once saw an interesting sight. A pair of ravens wanted to take possession of a heron's nest. The battle began on the tree in which the nest was built; but the ravens were driven from it, and then the fight was continued in the air. The ravens soared round and round, uttering their harsh notes, while the herons struck them with their sharp, strong beaks, and after a long time drove them off. In a heronry on the top of some noble Scotch fir-trees in Ashly Park, near Walton-on-Thames, a young heron had fallen out of its nest, and was brought into the house and given to a gentleman who happened to be there. It was put into a basket, which was placed in his gig, and he drove that evening to his house some miles distant. On arriving there, he turned the young bird into his garden, which was walled round. Early the next morning he saw an old heron feeding the young one, and this it continued to do until the latter was able to fly and make its escape. It would appear that this affectionate parent must have fled miles and miles in search of its offspring, until at last, hearing its plaintive cry, it came to its support.

The affection of birds for their young is very extraordinary. I have known a blackbird attack a cat that was near its nest while on the top of a wall, and by flying rapidly against it drive it away. This affection I have frequently seen in partridges and peewits, or plovers. When they have considered their young to be in danger, they will fly along the ground just before the person who is near their brood, flapping their wings as if they were wounded, and uttering piercing cries, thus drawing the intruder after them. It is a pleasing instance of maternal affection.

I do not know whether any of you remember an extraordinary flight of the small blue or rock-pigeon which took place over Brighton some years ago. I am informed that there was a similar one last year at this place. Where they come from, or where they go to, I am unable to inform you. In North America the flights of these birds are so enormous that they continue in one long, broad line for miles together, and towards evening they settle on trees in the forests, breaking down some of the branches, and many falling on the ground perfectly exhausted. The farmers in the neighbourhood know pretty well the time of this annual migration, and drive their pigs into the woods to feed and fatten on the pigeons.

But I must conclude, or I shall tire you. Let me, however, beg of you not to abandon your snug Fisherman's Home. When the advantages of it are more known, depend upon it a more general and liberal assistance will be afforded it by the inhabitants of this large and prosperous town. Some persons have said to me, "You will do no good amongst these fishermen—they will spend their money in drink and idleness as soon as they get it." Now, I think very differently, and far better, of you. I will never believe that the fine body of men I now see before me are incapable of receiving well-intended and kindly advice, and of acting upon it. I trust and think that this will not be the case.

I will only add, that in walking about Brighton I occasionally see in those vile receptacles for drunkards, called beer-shops, a paper stuck in the window with these words on it—"Best Old Tom here." Now, this Old Tom, as he is called, is a consummate rascal, as I am going to prove to you. You may ask, What has this old gentleman done to deserve such an epithet? You shall hear. If you form a too intimate acquaintance with him, he will lead you to poverty, misery, and ruin. He will make you impoverish your wives and children, and not only ruin your own health, but ruin also the health of your soul, rendering you at last a fit subject of the devil. But this is not all that this old rascal does. He has been even known to incite those who have been too much under his seductive influence to commit the crime of murder, and many, as they were ascending the fatal scaffold, have attributed their being led to do this from their having begun an acquaintance with Old Tom. Nor is this all. As a Magistrate for Middlesex, I can assure you that very many of those who are brought to our County Lunatic Asylum at Hanwell, in a state of insanity, both men and women, are indebted for their madness to this same Old Tom; for, I am sorry to say, that he has a very extensive acquaintance. What can I say more to you, except to urge and beseech every one to avoid his company and acquaintance, or, in other words, not to become a gin-drinker? From my long experience, I have known good, hard-working men, well to do, fall victims to this sad vice, and become ruined in this world and, it is to be feared, ruined in the next.

I have now done my best to amuse and instruct you. Do not think that what I have said respecting gin-drinkers is intended to apply to you. I see too many open, honest, healthy countenances before me not to feel sure that the vice of gin-drinking has been avoided by them. Farewell.

IV.

ON QUADRUPEDS.

My dear Friends,—

I have written three lectures for you, two on the subject of fish and one on birds, and I am now going to address you on the subject of quadrupeds, or four-footed beasts. I have a few anecdotes to tell you respecting them which I hope will interest you, but before I do this, I wish to impress upon you how desirable it is that you should become acquainted with the works of the Great Creator; and, be assured, that the more you acquire a knowledge of them, so much the more will you be led not only to admire but to wonder at the infinite variety and extraordinary contrivances of Him who made you and all things, both in heaven above and upon and in the earth beneath. If the most minute insect is examined through a magnifying glass, an exquisite and curious formation will be discovered, and will appear as wonderful as that of the largest animal. In short, it is our duty to see God in His works, and those works will declare His goodness.

And now I may tell you that much has been written on the subject of instinct and reason in animals, as well as in men. I will endeavour to explain the difference. Instinct leads all animals to do exactly what was first instilled in them at their creation. Birds build the same sort of nests and feed their young with the food most proper for them—the lion wanders about at night seeking his prey—the ostrich lays its eggs in the hot sand—the bee makes the same sort of curious cells—and so it is through all animated nature; but the dog, the elephant, and some other creatures, will sometimes act as if they were possessed of reason, and of which I will presently give you some instances. Man, on the contrary, is not led to act from instinct, but from reason. If you were going to commit an act of great folly or wickedness, reason would tell you not to do so, while instinct would teach a bird to avoid a hawk, and a rabbit to get into its hole to escape from a fox. I will now give you an instance of reason in men. You may remember the circumstance, for it only happened a very few years ago. A number of passengers, with many women and children, embarked in a ship, and in which was a company of soldiers. The ship sprung a leak, and it was soon evident that it could not be stopped, but that she must sink. The boats were lowered, but would not possibly hold the whole numbers on board. The brave, noble soldiers, called out that the women and children should be saved first; and they were so, the boats being soon filled with them and the passengers and crew. The soldiers were thus left alone on the deck, drawn up in line, dauntless and unwavering, their captain at their head. They felt the foundering vessel gradually sinking beneath them, but, strong in their sense of duty as well as of discipline, without an effort to save themselves, they went down with the vessel, and all perished; the women, children, and passengers and crew, were saved. Here was the triumph of reason over instinct. Instinct would have led these noble soldiers to seize one of the boats and to save themselves. Reason interposed, and triumphed, as you have seen, over instinct, and that in a way which did the greatest honour to British soldiers.

I will now give you some instances of reason in animals. Two friends of mine went out duck-shooting. When they came near some high reeds where they expected to find ducks, they threw their hats down, crawled to the reeds, and having shot at the birds, they sent their retriever dog for the hats, one of them being much smaller than the other. The dog took up first one hat in his mouth, and then trying to take up the second, the first, of course, dropped on the ground. After some efforts to take them both up at the same time, he put the smaller hat into the larger one, pressed it down with his foot, and then readily brought them both to his masters. This, I think, affords a strong proof of reason.

Another friend of mine was shooting on a hill in the north of England, which was surrounded by a stone wall, about four feet high. He fired at and wounded a hare, which ran through one of the holes left at the bottom of the wall. He sent his retriever after it, who readily leaped the wall from the higher ground, and pursued, caught, and killed the hare, and returned with it in his mouth to the wall. When there, he made some attempts to leap it, but failed. He then poked the hare through one of the holes with his nose as far as he could, jumped over the wall, went to the hole, and brought the hare to his master.

In Cumberland there are very extensive and high hills, on which numerous flocks of sheep depasture, and which at a distance look like little white specks. A shepherd will stand at the bottom of one of these hills, and send his dog up in the evening to collect his flock. This the dog will do by selecting the sheep from the different flocks, and bringing them down to his master, there being seldom one missing. Should there, however, happen to be one, the dog is sent back, and never fails to return with the proper sheep. I have watched this proceeding, and it has always appeared to me most wonderful that, in a flock consisting probably of some hundreds, mixed with several others, a poor dog should be able to distinguish each one of his master's sheep. A caress on the head, or a kind word, seemed sufficient to repay him for all his trouble. He will return at night to his master's cottage, wet and tired, and coil himself up before a fire, probably of a few sticks, and be ready to renew his toil the next day.

These sheep-dogs have a wonderful degree of intelligence. When I had a small farm, I was in the habit of having two hundred sheep sent me from the Cheviot Hills, some two hundred and fifty miles from my farm in Surrey. On asking the shepherd who brought them the first year how he had got on, he said he had but a young dog, and found much difficulty by the sheep taking wrong turnings, going up lanes and by-roads. The next year I asked him the same question. He told me that he had been accompanied by the same dog, who recollected all the false turnings the sheep had made the year before, and had gone before them and kept them in the proper road, so that he had no difficulty with them. Here was recollection, intellect, and a certain degree of reason as well as instinct.

The Highland shepherds are firmly convinced that their dogs perfectly understand what is said. Indeed, Hogg, the celebrated Ettrick Shepherd, related to me one or two instances in proof of this, which, I am sorry to say, I have forgotten;

but you shall hear another. A Highland shepherd, speaking to a gentleman, said accidentally, —" I'm thinking the coo (cow) is in the corn." His dog immediately rose, passed out of the house, and climbing to the top of a pigsty, which commanded a view of the corn-field, satisfied himself that the cow was not there, and returned to the house. In order to try the dog, he said, "'Deed, sir, the coo's in the taturs." Again the dog went out, made his own observations, and again returned. A third trial was then made, which showed that there was no occasion for the dog's services. He returned and went under the bed, sulky, growling, and dissatisfied, evidently disgusted at having been made a fool of.

A shepherd was in the habit of taking his little son with him, a boy of three or four years of age, when he was going to attend his sheep. He left him one day on the slope of a hill, while he went to some distance. On his return, he looked and hunted for the lad in every direction, but at last went back, late at night, to his cottage, and told his wife of their loss. While they were sitting together, miserable and disconsolate, they heard a scratching at the door.

On its being opened the shepherd's dog came in, which had not before been missed, and by his significant actions, by pulling the shepherd's coat and looking earnestly at him, induced him to follow the animal with his lantern, and was led by him to some rocks, into which the boy is supposed to have slipped, and thus the life of the child was saved.

I might multiply anecdotes of the sense of dogs to a great extent; but I will now tell you something of the sagacity of elephants, which, perhaps, have stronger reasoning powers than any other animal.

The father of a young lady who is now staying with me, was one day in a jungle in India tiger-shooting, mounted on the back of a favourite and much-petted elephant. All at once he saw a tiger crouching just beyond the head of the elephant. Having pulled the trigger, his rifle missed fire; he threw it on the ground in order to seize another, when, to his surprise, the elephant picked up the fallen gun with his trunk and gave it to him, as if aware that it was necessary for the destruction of the tiger.

Another day, this gentleman, while out tiger-shooting on the same elephant, was aware that a tiger was concealed in a very thick jungle close by, but from which he could not be driven. The mahout, or driver of the elephant, was desired to tell him to bend and beat the bushes with a tree which stood near. This the animal did so effectually, that the tiger started out and was shot. The gentleman was so pleased with the sagacity of his elephant, that he told the mahout to give him some sugar when they returned home—a favourite food with them. This was forgotten; but in the evening the sagacious animal found out his master, rubbed him gently and repeatedly with his trunk, and contrived to do so until the promised sugar was given him.

A third instance of sagacity and reason in the elephant I will now tell you. While on a shooting-party one of these animals got into a morass or bog. Nothing that it could do, or the attendants, could get it out. At last, some one suggested that a quantity of bushes should be thrown to it. These the sensible animal placed under its feet, and thus, by degrees, extricated itself.

An elephant, on one occasion, was ordered to drag a tree, which proved too heavy for its strength. It was urged and excited to continue the trial, till the poor animal broke the chains which attached it to the tree and ran away. It was supposed that it had escaped into the jungle, and would mix with the wild elephants. But how differently did this noble and sensible creature act! Instead of returning to its native wilds in the forest, it came back in about an hour, accompanied by two other elephants, and their united strength performed the task allotted to the first elephant. Here was reason and a power of communication in animals, which I have referred to in former lectures. You may doubt the accuracy of what I have stated, but I can assure you that the fact may be strictly relied on.

Let me tell you an anecdote of a seal, one of which was lately exhibited in Brighton; and a noble animal it was, and very obedient to its keeper. A gentleman, living near the sea in a remote part of Ireland, where the people are very superstitious, had a seal so tame, so affectionate, and so fond of its master, that it would follow and caress him whenever he had an opportunity of doing so. It so happened that there were two bad harvests in succession, and the foolish people attributed it to the poor innocent seal. They made such a stir about it that the proprietor was obliged to consent to its being sent away, provided its life was spared. It was placed in a boat, which was rowed to a considerable distance, and the seal then turned into the water; but it soon found its way back to its old master. A second time it was taken to a still farther distance; but again came back. A third time it was taken so far that the boatmen were absent two or three days; but, before they consigned this seal to the waters, they had the cruelty to put out its eyes. One day, the gentleman thought he heard the plaintive cry of his affectionate favourite. On opening his door, there was the seal, who had strength enough left to crawl so far, and then died: thus showing his love to the last. It must have died of starvation, as it was incapable of catching any food.

In the Firth of Forth, in Scotland, seals are very numerous, and will often put up their heads close to a boat. The fishermen, however, declare, that if there should be a gun in the boat no seal will ever come within its range. They are clever, sensible animals, and easily attracted by musical sounds, putting their heads out of water evidently for the purpose of listening to them. There is a well-known old seal in the Forth, who, from old age, has become perfectly white. The fishermen call him the Laird of Aberdour; and as they have never been able to kill him, they think that he cannot be killed. When a boat approaches the rock on which he is, he rolls himself into the water. The quantity of fish seals destroy is enormous, coming up the rivers after the salmon.

The beaver is another sagacious animal, living in companies, and acting together as if they were possessed of reason. In making their strong dams across the rivers in North America, they will, with their sharp teeth, gnaw through the bottom of a tree, so as to make it fall exactly where they want it. They then fill up the spaces with clay, to form the dam, using their flat tails, as bricklayers use trowels, in plastering it. In the recesses of this dam they lay up their winter store of food.

But I must conclude, with the hope that I may have amused you. I will endeavour to give you one more lecture before I leave Brighton, and then they shall all be printed, and sold for the benefit of the " Fisherman's Home," which I hope you will all stick to as a place of rational resort, and in the success of which I shall always take the greatest interest.

V.

ON INSECTS.

My dear Friends,—
One of our most celebrated poets has said—

> " Each crawling insect holds a rank
> Important in the plan of Him who framed
> This scale of beings;"

and it is on the subject of these insects that I am now about to address you.

The scale of beings in this world has been so beautifully and wisely ordained by the Great Creator, that one of the greatest of naturalists and philosophers in this or in any other country, while showing me a very small insect through his microscope, made the following remark:—" I believe that if it were possible to destroy the whole of these insects, the scale of created beings would become entirely disorganised or disarranged, so completely does the well-being of some depend on the existence of others." This is a curious and interesting remark, coming from such authority, and is well worth your recollection.

There are supposed to be about fifteen thousand different sorts of insects in the world; and many of them, if viewed through a microscope, would surprise you by the beauty and richness of their colours. Some have fins, like a fish; or a beak, resembling that of birds. Others have horns, like a bull or a stag. One is armed with tusks, not unlike those of an elephant; another has spines, like the quills of the porcupine or hedgehog; and some are covered with a substance like horn.

But perhaps the most interesting fact is, that there is no one invention of man up to the present time of which some hint may not have been taken from insects. You know that steamboats are made to pass along the water by means of a wheel on each side of the vessel. Well, there is an insect which moves itself in the water by the assistance of two little wheels fixed to its sides, which it turns round with great quickness, and thus goes from one place to another. Spiders may teach the art of weaving, and the bee that of building. One insect has an instrument like a saw; and another like an auger, or a carpenter's tool, to bore holes with. The nautilus, which may almost be called a sea-insect, spreads out a little sail, and guides itself with oars. In short, there is no creature, however insignificant it may appear, from which some benefit or instruction may not be derived.

You will, perhaps, be surprised when I tell you that many of the little insects you see around you have a great degree of sense, though you, probably, avoid them with dislike. But so it is, and I will give you one or two instances of it. You know that bees are kept in hives, where they lay up a store of honey, and they go in and out of the hive through a hole left open at the bottom of it. Now, a large slimy slug, which has no shell, got into a hive through this hole. The bees soon killed it; but their united strength could not drag it out of the hive, and therefore they covered it completely over with a thick coating of coarse wax, called *propolis*. It so happened that one of the common brown-shelled snails got into the same hive. It was soon stung to death; but, instead of covering it and its shell over with wax, they merely glued the edge of the latter to the board of the hive, and thus left it, as no unpleasant smell could issue from it, which would not have been the case with the slug had it not been cased over with wax. Was there not reason in this?

I had a hive of bees which was attacked by wasps, who wanted to get at their honey. In order to defend the entrance to the hive, they made a sort of fortification of wax behind the hole of entrance, leaving only two or three small passages just sufficient to enable one bee to pass at a time, so that they could defend themselves against the wasps with great ease.

When the heat on the inside of a hive is so great that the wax is in danger of melting, a number of bees will collect at the bottom of the hive, and move their little wings with such rapidity, that they can create as great a circulation of air as a lady would do in fanning herself.

In very hot countries, such as the West Indies, bees alter their mode of laying up their store of honey. If they placed it in cells at the top of the hive, as they do in colder countries, the great heat of the sun would melt the wax, and all their fanning would be to no purpose. They therefore place it in very small waxen bottles, somewhat in shape of an inverted mushroom, the stalk or neck being uppermost. These are placed at the bottom of the hive, and in this situation the wax is kept cooler, and, of course, less liable to suffer from heat.

I think that these facts will show you that bees are possessed of sense, or what some would call a superior intellect. It is not, however, only their sense which we should admire, but also their industry and usefulness. They are never idle, but work from morning to night in collecting honey; and when the weather is bad they employ themselves in cleaning the hive, and in making and repairing their cells. These cells are beautifully constructed; and, as the bees are obliged to pass over them very frequently, the edges of them are much stronger and thicker than the walls of the cells. These edges, however, serve another purpose, as they help to retain the honey. When the cells are filled, they are covered over with wax, and not opened until the honey is wanted.

Ants are another class of insects whose operations are curious and wonderful. For instance, there is a small red ant in the West Indies which conceals itself in covered ways, and attacks and feeds on the hardest woods, never appearing on the bark, or even touching it. Thus, when a tree or a beam of a house appears perfectly sound, it has been, perhaps, eaten out, and nothing but a shell remains.

Some of our English ants will ascend a poplar or a lime-tree, where they find, on the tender shoots of the tree, a small green insect called an aphis. You may ask, "Do they eat these insects?" Quite the contrary. The ants, as I have often seen them do, tickle them with their antennæ, or little feelers, which project from their heads. This seems to please the aphides, who discharge a sweet substance from their bodies, called honey-dew, on which the ants feed. They may be called the milch-cows of the ants.

In a fine morning, ants disperse themselves in various directions and to considerable distances from their usual abodes in search of food. When they have discovered any, they make their companions acquainted with it by means of their antennæ, or feelers, which serve them, as I will presently prove to you, to hold a sort of conversation with each other. You may be surprised at hearing this; but it is perfectly true, not only with respect to ants, but also with bees and wasps, and probably, also, in regard to other insects. I have frequently placed a small green caterpillar near an ants' nest, and watched what would take place. A solitary ant has, perhaps, discovered it, and eagerly attempted to draw it away, as a winter store—for they lay up, as prudent persons should do, against a rainy day. Not being able to accomplish this, I have seen it go up to another ant, and by means of the antennal language bring it to the caterpillar. Still, these two were not able to accomplish the task. They separated, and brought up reinforcements of their community by the same means, until a sufficient number were collected to enable them to drag the caterpillar to their nest. You thus see that a sort of language can be kept up by means of these antennæ. It is supposed that a strong hive of bees will contain 36,000 workers. Now, each one of these, in order to be aware of the presence of their queen,—whom they love, as I am sure you do yours,—touch her every day with their antennæ. Should the queen die, or be removed, such is the affection of her subjects for her, that the whole colony disperse themselves, and are seen in the hive no more— perishing, every one of them, and quitting all their store of honey which they had laboured so industriously to collect. On the contrary, should the queen be put into a very small wire cage, placed at the bottom of the hive, so that her subjects could touch and feed her, they are perfectly contented, and the business of the hive proceeds as usual.

You see, then, that this antennal language is a wonderful and curious gift, bestowed by a Benevolent Creator on little insects; but who are all, like you, objects of His love and care. You should always bear this in mind; for, if God clothes the flowers of the field, and feeds the young ravens that call upon Him, be sure that He will both clothe and feed you if you trust in him, and endeavour to do what is pleasing to Him.

There is another class of insect, the common house-fly, well worth your notice. On examining them, you will see that each has two projecting eyes, both of which are furnished with 4000 lenses, appearing, through a microscope, like a piece of honeycomb. These give them that rapidity of motion which you must often have observed in escaping capture. Their flight, also, is so swift that it has been supposed they can fly at the rate of nearly a mile in a minute. Their interior structure is equally curious, having blood-vessels and other functions which are to be found in larger animals. But I must draw your attention to their feet, which are furnished with a sort of wet sponge, that enables them to run up and down smooth surfaces, such as glass, with the greatest facility. Some flies lay their eggs in dung-pits; others in the bark of trees, which causes them to throw out those gnarled and wooden projections which you may sometimes see in old oak and elm-trees.

Wasps are another species of insects whose proceedings are very curious and interesting. Early in the spring, and on a sunny day, you may see a large wasp settling on some decayed wood and appearing to feed on it. This is a female wasp, the parent of a large colony. The wood she seems to eat is formed into a few cells, perhaps at first only five or six, in which she deposits her eggs. These are gradually increased in number until the eggs are turned into wasps, which then come out and assist in the work of building, until what may be called a large city is formed, inhabited by some thousands of her subjects,—for the queen is the mother of them all. She generally selects a hole in a bank, some old hollow tree, or a roof of a house, in which to make her future colony. There are, however, two distinct species of wasps in this country; one of which forms her nest, generally a smaller one, on the boughs of fir and other trees. They are round, and are covered with what may be called flakes, of a substance made also from wood. These throw off the rain, and also serve to keep the young wasps warm before they arrive at maturity. The other species of wasps make the same covering.

In order to show you with what rapidity these insects form their nests, I will mention the following fact:—I built a small shed early one spring, which was covered over with slates. In the course of the following autumn it was found that the slates admitted the rain. Some of them were, in consequence, taken off, when a wasps' nest was found built on the rafters, so that it could readily be removed perfectly entire. It was a beautiful structure, the outside appearing as if covered with small brown shells, and it measured $3\frac{1}{2}$ft. in circumference. I sent it to the Zoological Museum. Now, this large nest must have been made in the course of five or six months.

The nests of hornets are equally curious; but they use green wood in forming them. They are generally found in hollow trees, but sometimes on the branches of firs; so that, probably, like the wasps, there are two species. They are a dangerous insect, as their sting is very severe. They are also very strong. I once placed a hornet under a common wine glass, when it got its fore feet under the edge of the glass, and thus lifted it up, and made its escape. I also once saw a hornet carry away a small pear from my garden. Like wasps, the female hornet is the parent of the colony. She remains torpid during the winter, and in the spring leaves her retreat, and begins making her nest. All the workers, both those of the wasps and hornets, die in the winter. This is a beautiful arrangement of Providence, or, rather, I should say, of the Great Creator; for if these insects increased and multiplied in the same proportion as bees, we should experience a most unbearable nuisance. Bees, on the contrary, increase; are confined in hives; and are most beneficial to mankind in producing wax and honey. I had, however, almost forgotten to tell you that there is a winged ant in South America which builds a curious and beautiful nest in trees. The cells of the combs are hexagonal, like those of bees, and are stocked with honey, which is both good and sweet. The cells are much smaller than those of bees, and the whole arrangement of them shows the most interesting specimen I have seen of insect architecture. I have brought a piece of their comb to show you.

The large forests in South America would be almost impassable, both for man and beast, if it were not for the beetles. The trees which compose them stand very thick, and as they and the branches are constantly falling down, they are quickly consumed by beetles, which abound in these forests in incredible numbers. Ants also assist in the destruction of the trees, so that the forest can be travelled through without many impediments. But for these useful insects, no one could get through them.

The nests of many insects, as I have mentioned, are very curious and interesting. I have brought two or three specimens of them to show you. One is that of a moth, found in South America. You will see with what labour and wonderful skill it is formed, and how impossible it seems for an insect, shaped as a moth, to introduce these numerous bits of sticks and fix them on the outside of its nest. Here are some other specimens I have brought to show you.

But I must not forget to tell you of a little creature which lives far out at sea, and is found on the gulf-weed. Its name is latiopa. Sometimes a rough wave will sweep it from the weed and force it into the deep waters; but it is provided with an air-bubble, and it glues to this bubble a thread, which it lengthens as the bubble naturally rises to the surface. This small quantity of air, before it bursts, floats on the water, and is soon attracted by the gulf-weed, towards which it runs and fastens itself alongside. Then up comes the insect, by means of her thread, and thus regains her seat on the weed,—her natural position. You see how wonderfully and kindly the Great Creator has provided for the well-being of a little insignificant creature.

There are, no doubt, many curious and interesting insects in the sea; but, from the element in which they live, we know but little of their habits and peculiar instincts. Some are parasitic, or adhere to the bodies of fish, and others eat into the flesh of whales. I must, however, refer to the mussels, which, as you are aware, attach themselves to rocks, by means of a strong silk-like thread; but perhaps you are not aware that these threads, which are very strong, have been collected and made into a pair of gloves.

I have deferred my mention of the most interesting and valuable of insects for the conclusion of my lecture. I refer to the silk-worm. In order the better to enable you to understand the wonderful arrangements of Providence in regard to this moth, I may tell you that all butterflies and moths, when they quit their cocoons, or winter-coverings, fly away, and are of no service to man, except as objects of admiration of their beauty and peculiar instincts. But how different is the case of the silk-worm! If she were furnished with expanding wings, like other moths, she would fly away and become useless. Her wings, however, are so short, that although she flutters them a little, she never attempts to quit her home which has been provided for her, but lays her eggs and dies—so short is her existence as a perfect insect. These eggs, when hatched, produce very small grubs, which must have mulberry leaves provided for them to feed upon. When they have arrived at maturity they leave off eating, and begin to spin their valuable silk, forming it into a cocoon, or covering, which they may be heard doing night and day, for three days, when it is finished. The silk which covers the cocoon will extend for many hundred yards. In the silk countries—that is, in places where the silk-worms are most cultivated—these cocoons are thrown into a cauldron of hot water; the ends of the silk unfasten, when 50 or 60 of them are caught and wound off into hanks, or skeins, like worsted, all together, and sold by weight to silk merchants. The grubs which spin these cocoons, of course, are killed; but a certain number are reserved for breeding, and these lay many eggs.

Now, let us consider the utility of the silkworm. The countries which produce the greatest quantity of silk are Italy, Turkey, India, and China. The number of persons in the world who are benefited or clothed by silk is perfectly enormous. There are the growers of it, the feeders, the cultivators of mulberry-trees, the manufacturers of silk, the shopkeepers and their attendants, and many others who are either directly or indirectly benefited by it; and all this is owing to the peculiar construction of an apparently insignificant insect. And here I may mention that no fishermen's knots are found in winding off the silk, and no entanglement or breakage, but all runs smoothly from one end to the other. Let us wonder at and admire this wonderful arrangement of a Benevolent Creator in forming this insect.

But I must conclude this, my last lecture— at least, for the present— with every kind wish for your happiness and prosperity. Do not forget your old friend, but try and make a good use of what he has said to and written for you, and God bless you all!

VI.

THE LOVE OF ANIMALS FOR MAN AND FOR EACH OTHER.

My dear Friends,—

Some of you asked me to read you another lecture, and I am going to give you one on the love of animals for man and for each other. It is an interesting subject, and may, perhaps, induce those who either read or hear it to treat dumb animals with that kindness which every one who has a good heart would wish to do.

When that fearful curse was pronounced upon man, "the fear of you and the dread of you shall be upon every beast of the field and fowl of the air," leading creatures to avoid mankind as their worst enemies, one exception seems to have been made in the case of the dog. This faithful animal cleaves to his master through poverty, distress, hunger, and even death itself. Nothing destroys his love and attachment. We have instances when officers have been killed in battle, a loving dog has remained close to the body of his master, howling his distress, refusing all food and comfort, following the body to the grave, and expiring upon it,—thus showing his affection to the last.

Another affecting instance is related of a dog, who followed his master to his grave, which was in one of the London churchyards, and was overlooked by several houses. On this grave the dog scratched a hole and lay in it. One of the kind-hearted inhabitants of the houses brought it some food; but there it remained day after day, but eating what was brought it. At last, some one erected a small shed over it, to shelter it from the weather. There the dog might be seen year after year, protected and sympathised with by every one who knew the circumstance of its remarkable attachment, till death — and nothing but death — released it from its fidelity to the master it loved.

But, before I proceed with my anecdotes of these affectionate animals, I must express my surprise that so many unfeeling allusions should constantly be made to those noble creatures. Thus we hear of a "lazy dog"—a "drunken dog"—a "dirty dog"—a "shabby dog"— of leading a "dog's life"—of a "dogged temper." We call a dandy a "puppy," and sometimes another man a "cur." All these are epithets misapplied as far as the dog is concerned; and I think you will agree with me when you hear the following anecdote:—

A young gentleman of the name of Gough, of considerable talents and of an amiable disposition, lost his way when wandering, without a guide, on the mountain Hellvellyn, in Cumberland. He was accompanied by a terrier bitch, his constant attendant during frequent solitary rambles through the wilds of Westmoreland and Cumberland. Trying to reach the top of the mountain by a difficult pass, he fell down a precipice called "Stridenedge," and was killed. His remains were not discovered until three months afterwards, when they were found guarded by his faithful dog. Although the body had been so long a time exposed to the attacks of the numerous wild birds of prey, and also the foxes which abound in that region, it was found untouched and undefaced by them, so strictly had it been watched over and protected. How the dog procured his food is a mystery which has never been discovered, but the fact of his remaining near the body of his master is undoubted. The celebrated Sir Walter Scott, a great lover of dogs, and who frequently visited a friend in Cumberland, wrote a beautiful poem, called " Hellvellyn," on the incident I have just related. It is too long to quote the whole of it, but I will give you an extract from it. Sir Walter Scott says he had seen the place where the wanderer had died, and then adds:—

" Dark green was the spot mid the brown mountain-heather,
   Where the Pilgrim of Nature lay stretch'd in decay,
Like the corpse of an outcast abandon'd to weather,
   'Till the mountain-winds wasted the tenantless clay;
Nor yet quite deserted, though lonely extended,
For, faithful in death, his mute fav'rite attended,
The much-lov'd remains of her master defended,
   And chased the hill-fox and the raven away.
How long didst thou think that his silence was slumber?
   When the wind wav'd his garment, how oft didst thou start?
How many long days and long weeks didst thou slumber,
   Ere he faded before thee, the friend of thy heart?

And, oh! was it meet that—no requiem read o'er him, .
No mother to weep, no friend to deplore him,
And thou, little guardian, alone stretched before him—
Unheeded the Pilgrim from life should depart?"

I am sure you will all thank me for relating this beautiful and affecting anecdote of the love and fidelity of a poor dumb animal to his master, whose remains were interred in the burial-ground attached to a Quakers' meeting-house, near the foot of the mountain.

A poor woman, returning one winter's evening from a market, where she had purchased her loaf of bread, a bit of bacon, and a small piece of mutton, and accompanied by a small dog, was overtaken by a violent snow-storm, as she was passing along a narrow lane. She was unable to proceed, and at the end of three or four days was found dead. Her dog had survived, and was discovered close to his mistress and the basket of food, which was untouched, although the poor animal must have been nearly starved from having had nothing to eat for so long a time.

I will now relate another anecdote of the love and affection in dumb animals, which I am sure will please you. Mr. Morritt, well known to the readers of the life of the celebrated Sir Walter Scott, as his intimate and confidential friend, had two terriers of the pepper-and-mustard breed, or, rather, for it is a character I always delight in, the Dandy Dinmont breed. These dogs,—for it is as well to leave out the feminine appellation,—were much attached to their kind-hearted master, and he to them. They were mother and daughter, and each produced a litter of puppies about the same time. Mr. Morritt was seriously ill at this period, and confined to his bed. Fond as these dogs were of their puppies, they had an equal affection for their master, and they accordingly showed this in the following manner. They conveyed their two litters of puppies to one place, and while one of the mothers remained to suckle and take care of them, the other went into Mr. Morritt's bedroom, and continued there from morning until evening. When the evening arrived, she went and relieved the other dog, who then came into the bedroom, and remained quietly all night by the side of the bed; and this they continued to do, day after day in succession, until Mr. Morritt recovered. This charming anecdote was communicated to me from a quarter which need not leave a doubt of its authenticity, and affords an affecting proof of love and gratitude in animals who, I am sorry to say, are too often ill-treated.

A vessel was driven by a storm on the beach of Lydd, in Kent, a place some of you are probably acquainted with. The surf was rolling furiously, and eight men were calling for help; but not a boat could be got off to their assistance; although I have no doubt but that some of you now present would have tried, for the Brighton fishermen have done many brave things. However, a gentleman at length came on the beach, accompanied by his Newfoundland dog. He directed the attention of the noble animal to the vessel, and put a short stick into his mouth. The dog at once understood his meaning, and sprang into the sea, fighting his way through the foaming waves. He could not, however, get close enough to the vessel to deliver the stick he was charged with, but the crew joyfully made fast a rope to another piece of wood and threw it towards him. The sagacious dog saw the whole business in an instant—he dropped his own piece, and immediately seized that which had been cast to him, and then, with a degree of strength and determination almost incredible, he dragged it through the surge and delivered it to his master. In this way a line of communication was formed, and every man on board saved. Does not this anecdote make you love dogs? It ought to do so.

I will now give you an instance or two of the love and kindness of animals to each other. My home is at East Sheen, in Surrey; but a short distance from it there resides an amiable and excellent gentleman, who, like many others, has his cows, pigs, and poultry, and one of his pigs produced a large litter. As is generally the case, the youngest of the litter was a small, weakly pig, and was pushed away when he attempted to feed with the others. Being thus without food, he gave utterance to his plaintive, feeble cries. These attracted the sympathy of a kind-hearted hen in the yard, who sheltered and warmed it under her wings. The pig was subsequently fed by artificial means, but the hen continued her care of it till it no longer required her protection.

I will give you another instance of animal

kindness which occurred under my own observation. The late Earl of Albemarle, when Master of the Horse to the Queen, lived at the Stud House in Hampton Court Park. He had a fine breed of black-and-tan spaniels, one of which produced a litter of puppies and died in bringing them forth. Their plaintive cries, like those of the pig I have mentioned, induced a young female of the same breed, who never had puppies of her own, or was in the way of having any, to foster and warm them. They attempted to suckle her; milk came in consequence, and she was thus enabled to bring them all up. I have often seen her employed at her task, and nothing could exceed the affectionate way in which she performed it.

You shall hear of an instance of friendship in animals. When the German Legion was actively engaged in the Peninsular War, two horses were always picketed together and served side by side in the same troop. One of these horses at last died. His companion refused all food, and pined away and expired — a victim to his affection for his constant companion. Birds, also, that have been kept together in cages, have been known to die when they have lost a companion: so capable are animals of showing love and affection.

But let me return to the dog; for I like to dwell on his noble qualities. It was a pleasing remark of Sir Edwin Landseer, whose pictures of dogs approach so near to the life, that the Newfoundland dog was "a distinguished member of the Humane Society." Indeed, we see in Sir Edwin's pictures faithfully portrayed honesty, fidelity, courage, and sense—no exaggeration—no flattery. He makes us feel that his dogs will love us without selfishness, and defend us at the risk of their own lives; that though friends may forsake us, they never will; and that in misfortune, poverty, and death, their affection will be unchanged and their gratitude unceasing.

A gentleman, while bathing in the sea near Portsmouth, was in the greatest danger of being drowned. Assistance was loudly called for; but no boat was ready, and, although many persons were looking on, no one could be found to go to his help. In this predicament a Newfoundland dog rushed of his own accord into the sea, and was the means of saving the life of the gentleman. He afterwards purchased the dog for a large sum, treated it as long as he lived with great kindness, and had the following words worked on his table-cloths and napkins—" *Virum extuli mari ;*" which may be thus translated—" I have rescued a man from the sea."

You will be amused with the following anecdote, for it is something in your way as sailors. There was a Newfoundland dog on board H.M.S. Bellona, which not only kept the deck at the bloody battle of Copenhagen, but ran backwards and forwards with so much courage and apparent anger at the foes, that he became a greater favourite than ever with the crew. When the ship was paid off, after the Peace of Amiens, the sailors had a parting dinner on shore. Victor, the dog, was placed in the chair, and fed with roast beef and plum-pudding, his health drank, and the bill made out in Victor's name.

A kitten, only a few days old, had been put into a pail of water in the stable-yard of an inn for the purpose of drowning it. It had remained there for a minute or two, until it was to all appearance dead, when a female terrier, attached to the stables, took the kitten from the water and carried it off in her mouth. She suckled and watched over it with great care, and it lived and thrived. She had at the time a puppy of her own.

I will now tell you something of the turnspit. In my very young days I was at a school where large joints of meat were turned by two of these dogs, one on one day and the other on the next. When you consider that a joint of beef would take at least three hours to roast, you may suppose that the poor dogs had no easy task to perform. The consequence was that, as the dinner-hour drew near, they would often hide themselves, and I have been told that if one of them was found, and it was not his turn to be put into the wheel, he would point out the retreat of his companion, showing that they not only calculated time, but are clever, sensible dogs.

Dogs have been known to die from excess of joy at seeing their masters after a long absence. An English officer had a large dog, which he left with his family in England while he accompanied an expedition to America during the wars of the colonies. Throughout his absence the animal appeared very much dejected. When the officer returned home, the dog, who happened to be lying at the door of an apartment into which his master was about to enter, immediately recognised him, leaped upon his neck, licked his face, and, in a few minutes, fell dead at his feet. A favourite spaniel of a lady recently died on seeing his beloved mistress after a long absence.

I have now given you some anecdotes of the affection, sense, and strong attachment of animals either to man or to each other, especially of the dog. I am convinced that, the more the character of this animal is known, the better treatment he will receive, and a stronger sympathy will be excited for him. In fact, he is a friend so faithful — a protector so disinterested and courageous — that he deserves all the kindness and affection which can be shown him. A French writer has boldly affirmed that, with the exception of women, there is nothing on earth so agreeable, or so necessary to the comfort of man, as a dog. However this may be, it is certain that if we were deprived of the companionship and the services of the dog, man would be a solitary, and, in many respects, a helpless being. The dog has died in defence of his master — saved him from drowning — warned him of approaching danger, and, if deprived of sight, has gently and faithfully led him about. If his master wants amusement in the fields or the woods, he is delighted to have an opportunity of procuring it for him. If he finds himself in solitude, his dog will be a cheerful companion; and maybe, when death comes, he will be the last, as we have seen, to forsake the grave of his beloved master. In fact, he is fond, intelligent, and grateful. I will here quote some lines by Lord Byron on his dog:—

> " When some proud son of man returns to earth,
> Unknown to glory but upheld by birth,
> The sculptor's art exhausts the pomp of wo,
> And storied urns record who rests below:
> Not what he was, but what he should have been.
> But the poor dog, in life the firmest friend,
> The first to welcome, foremost to defend,
> Whose honest heart is still his master's own,
> Who labours, fights, lives, breathes, for him alone,
> Unhonour'd falls.   *   *   *
>   Ye who perchance behold this simple urn,
> Pass on—it honours none you wish to mourn;
> To mark a friend's remains these stones arise—
> I knew but one, and here he lies."

I have now done with my lectures for the present. You wished, when I last met you, to have one more, and I need not tell you that this has been a hurried one. I have only to say that I quit you with much regret; for I have been listened to with kindness and attention, and, what has gratified me much more, I have been assured that I have both pleased and instructed you. Be quite certain that I shall never forget you, or cease to feel a deep interest in your welfare. Do not forsake your "Home." It is a place for rational enjoyment and improvement, equally a place also for an honest, sober man, as well as a Christian. And now farewell.

VII.

THE INFLUENCE OF ANIMAL LIFE ON LIME.

My dear Friends,—

I propose in this lecture to talk to you a little on those sea-flowers, which, on a calm sunny day, you have seen floating about, expanding their beautiful colours, and unfolding their living petals to entrap their food. I propose also to notice sponges, corallines, and other marine productions, and to point out to you the use which Divine Wisdom intended they should be of to you fishermen when He created them. You may ask, How can these apparently almost inanimate substances be of use to us? I will endeavour to explain this as intelligibly and plainly as I can, but at the same time expressing my belief that without them your occupation as fishermen would be a useless one—

that is, that you would not be able to take fish in your nets.

Now the sea, as you know, consists of a vast expanse of water, and Providence has many methods of keeping it in a pure and wholesome state. It is impregnated with salt—the winds blow over and agitate it; and now let us see the effect of a storm upon it. The sun is obscured, and the breeze freshens from the sea—dark clouds are gathering on the horizon, and the tide begins to turn. The heaving waves now tumble towards the shore, and, as they break in angry foam, portend a storm. The sky looks threatening, and the thunder peals in the distance. The sea appears to wake as from a slumber, and the blackening heavens lower over its dark bosom, while the rising blast, impelling all her waves, drives them upon the rocks in sheets of white foam, lashing them on as if to madness, till at length ocean and sky seem mingled, and all is violence, and roar, and rage.

Such are the changing aspects of the sea, and such the efficient means whereby Nature insures her renovation of the mighty deep, refreshing it throughout its broad domains, and keeping its waters wholesome, filled with air, and thus adapted to afford the means of respiration to the numberless living things that flourish in its vast space.

But there is another operating cause, which I have already hinted at, and most wonderful it is, in rendering the waters of the sea conducive to the well-being of the animals which are to be found in it, and which I will now do my best to explain to you. You are aware that all rivers discharge themselves into the sea, and these waters have a strong impregnation of lime, which is obnoxious to animal life. Now, let us see how this is obviated. It is a curious and interesting fact, and I have the authority of the first naturalist and philosopher of this, or, perhaps, any other country (Professor Owen), for what I am going to tell you. You have all of you probably heard of the coral insect—a very small one, indeed. By the agency of these coral animalculæ, two hurtful influences of the caustic lime are neutralized. These animalculæ absorb the lime which is in solution, combine it with a substance called carbonic acid, and thus form what is called carbonate of lime, which is insoluble, and perfectly innocuous, or not detrimental to animal life. By thus forming and precipitating, or forming at the bottom as a sediment, this earthy matter, they are enabled to build up the cells in which they live, and by which their soft, jelly-like substance is defended. These cells form what is called the coral or madrepore, and the quantity of caustic lime thus removed from its state of solution in sea-water, and thrown down or precipitated as an insoluble earth, or earth gathering a crust, may be conceived when I tell you that islands and reefs of coral, upwards of one thousand miles in extent, are formed by the agency of these seemingly frail and insignificant living beings.

I have said that these insects form islands, and, before I proceed further to show you the object of animal life in removing the caustic lime from the sea, I will tell you how this is performed. The operations of the coral insect, through a long succession of ages, proceed upwards from the original foundations, until the surface of the sea is reached, when the work ceases. Sea-birds settle on these rocks of coral, sea-weeds are driven on them, and other substances, all of which, added to the manure of birds, form a stratum of soil, on which seeds are either dropped or washed by the tide, and thus take root and flourish, so that in time an island is formed, covered with cocoa-palms and other exotic trees. This is the case about the Bermudas, where nothing can exceed the beauty of these coral islands.

When you consider that, in addition to these, there are coral reefs, some of which extend upwards of one thousand miles, it would appear certain that the operations of these little insects must have existed from the earliest periods of this our globe. It is a wonderful fact to reflect upon, and will show how Providence arranges beneficially for the good of His creatures a system which it would never have entered into the mind of a finite being to conceive as possible, until more recent discoveries have proved that such is the case. Barrier reefs are similar to coral islands. They run parallel with the shores of some larger island or continent, separated, however, from the mainland by a broad and deep lagoon channel. The prodigious extent of the combined and ceaseless labours of these little world-architects, as they may be called, must be witnessed in order to be properly conceived. For instance, they have built up a barrier reef along the shores of New Caledonia for a length of four hundred miles, and another which runs along the north-east coast of Australia one thousand miles in extent. Now, supposing this latter to be only a quarter of a mile in breadth and one hundred and fifty feet deep, here is a work compared with which the walls of Babylon, the Great Wall of China, or the Pyramids of Egypt, may be called children's toys. You must recollect also that the operations of the coral insect were carried on amidst the waves of the ocean, and in defiance of its storms, which, as you know, often sweep away the solid works of man.

Such are some of the operations of that extraordinary insect, the coral; and I have pointed out their influence on the caustic lime which finds its way into the sea, and which is so injurious to animal life. But there are other means provided for neutralizing and rendering the lime harmless. Sponges, sea-flowers of various descriptions and in innumerable quantities, and other marine animals, all having life, act as agents in this wise arrangement of Providence to purify the waters of the ocean. Sponges, you might suppose, have no vitality or living organs; but this is not the case, although they perhaps are the lowest in the scale of living beings, as I will show you. The sponge, as you know, not having the power of moving from place to place, is fixed and motionless as the rock on which it grows. Now, as the sponge is productive, that is, that it propagates other sponges, it may be asked by what means this is done? The parent sponge produces a large quantity of seeds, called gemmules, which may be compared to very small pins' heads. If these are put in a watch-glass containing a little sea-water, it may be seen by the naked eye that they are able to swim about with great facility and quickness. On examining them with a good microscope, the way in which they move will be discovered, and certainly a wonderful sight it is. Millions of what may be called paddles, and all furiously at work, cover the surface of these tiny atoms, and all so rapid in their motion that it is almost impossible to perceive their shape. These living seeds are washed out of the body of the parent sponge by the currents of the sea, and in process of time fix themselves on the rocks and there grow into sponges, and fulfil at last the purposes for which nature intended them, that of neutralizing the effect of caustic lime; and if this were not done, very few fish would be taken by you fishermen. You see, then, how wonderfully Providence, or rather an Almighty Being, has arranged everything, not only for the good of His creatures, but for mankind generally; and this by the means, in some cases, of little insignificant insects. And here let me ask you to pause for one moment in order to contemplate the amazing power and wonderful goodness of Him who made the earth and the sea, and stored them both with what is useful and beneficial to us. Then look at the sun, the moon, the stars, and the glorious expanse of the heavens, and see what power is displayed in them! Gladly would I call upon you to look up with reverence to that Great Being who is so good, so wise, so considerate to all His creatures, and to pray to and love Him as He directs us to do. Great God! what sacrifice is too great to offer Thee? what love too devoted? Indeed, His love to us—His goodness to us—His care of us, all demand this tribute from us, and hard and insensible must our hearts be if we do not show it.

But to return to the benefits which are derived from the innumerable small creatures which are found in the sea. Independently of what has been stated of their removing the noxious influence of caustic lime, they are also agents in the hands of the Almighty for other purposes. Take up a handful of the sand cast up by the retreating tide upon the adjoining beach, and with a magnifying glass examine it minutely; count the various objects you will see there, and if you have sufficient patience, count their numbers also: but it will be no easy task. You will see how innumerable are the forms in which very small shells will present themselves, all of them delicate in their structure and elegant in their shape, and inhabited by exquisitely-formed creatures, having all the functions of those found in larger shells. Monsieur D'Orbigny, a great observer of these minute shells, reckoned 3,840,000 in one ounce of sand! If we attempt to calculate the contents of a square yard, the amount would surpass all human conception. So incredible are their numbers that they form banks, which, by their accumulation, interrupt the progress of ships, stop up bays and straits of the sea, fill up harbours, and with corals produce those islands that rise up in the Pacific Ocean. But do not start when I tell you that these almost invisible living objects have been employed in building up the very world on which we tread. I should be sorry to state anything to you without some proof of what I said. It is easy, therefore, to show you that these almost invisible shells have had to do with the construction of the surface of the earth. Take, for example, the neighbourhood of Paris. The chalky substance about it is in some places so filled with these shells, that a square inch from one of the quarries contains something like 58,000 of them, and that in beds of great thickness and of vast extent. This would give an average of many millions of millions in the square yard. Now, as all Paris and the towns and villages of the neighbouring districts are built of the stone quarried from this deposit, it is evident, without any exaggeration, that the capital of France and all the neighbouring towns are constructed principally of these little shells. Indeed, the chalk of our own dear country, " the white cliffs of Albion," which you all so well know, and which throughout is of such vast thickness, contains myriads upon myriads of these interesting shells.

But I shall tire you if I proceed further on this subject. Be quite sure that I have the authority of some of our best naturalists for what I have stated; but your own eyes and examinations may help to convince you that at least a part of what I have said is true. One of my objects has been to amuse you—the other to raise your thoughts to that stupendous Power, who, apparently by small means, works such great ends, and who is at the same time infinitely good, wise, and loving to every one. In this world you are always in danger of foundering amidst the waves of sin, or of being shipwrecked on the rocks of temptation; but, like wary pilots, steer your boats into some well-sheltered creek till the storms of this world are blown over, and enjoy a safe anchorage. Then hoist your flag of hope, ride before the sweet gale of redeeming love, till you make, with all the sails of humble faith, the blessed port of eternal life and happiness.

VIII.

ON
INSECTS AND MARINE ANIMALS.

My dear Friends,—

You were pleased with my lecture on Insects last spring, and I am now going to pursue the subject; for it is almost an endless one. In fact, the works of nature, or, I should say, of the great Creator, far exceed what we know, or, indeed, are ever likely to know of them. My lecture, however, will not be confined to insects; but I propose also to bring under your notice some of those marine animals which so many of you must have seen, but, probably, have not been acquainted with their peculiar habits.

I will first describe to you the cochineal insect, because, with the exception perhaps of indigo, it produces the most important of all materials for the dyers, and because I omitted to mention it in my first lecture on Insects. The cochineal is extensively cultivated in Mexico, and when the Spaniards conquered that country in 1518, they found the fine dye procured from this insect, but supposed for a long time afterwards that it came from a very small seed; for such they thought it to be in consequence of its singular appearance. It was only from its being examined under a microscope that its true nature was ascertained. The insect, when imported into this country to be used as a dye, has the appearance of a reddish shrivelled grain covered with white powder. It feeds on what is called the Indian fig, a species of cactus, commonly called the prickly pear; but in Mexico, where the cochineal is most cultivated, the hopal. In that country, plantations of the hopal may be seen in lines of fifty or sixty thousand, each plant being kept about four feet high for more easy access in collecting the insect. Great care is required in this operation, which is performed by the Indian women with a squirrel's or stag's tail. When the insects have been collected they are killed, either by throwing them into boiling water or by placing them in ovens; and these latter bear a higher price in the market, as they are less subject to adulteration, a whitish powder being preserved on them. The quantity of these insects imported from South America for the purpose of procuring a beautiful red dye amounts to upwards of half-a-million of pounds sterling,—a great sum to be derived from so small an insect, and which should show us the folly of despising any animal on account of its apparent insignificance and minuteness. The Spanish Government have always been extremely jealous of any interference with their trade in this insect: the East India Company offered a reward of 5000*l.* to any one who should introduce it into India; but the experiment has, I believe, hitherto failed.

Another insect possesses that valuable material called lac, and is a species of cochineal. It is found on various trees in the East Indies, and the substance is made use of in that country in the manufacture of beads, rings, and other female ornaments. When added to sand, it forms grind-stones; mixed with lamp or ivory black, being first dissolved in water, and with a little borax, it makes admirable ink. A new preparation of lac-dye is now used, and which, when mixed with cochineal, makes so fine a scarlet that it is said to have saved the East India Company 14,000*l.* a-year in the purchase of scarlet cloth for the army.

The quantity of wax made by bees is enormous. In Spain alone, one single parish priest is said to have had five thousand hives. In Russia many peasants have four or five hundred bee-hives, and make more profit of their bees than of their corn. Since the introduction of sugar, honey has lost much of its importance, but still it is of great value in this country.

I have now endeavoured to point out to you some of the benefits which are derived from insects, and the knowledge of this cannot, I hope, have failed to impress on your minds the wonderful love of the Great Creator, who has formed everything for some good and wise purpose, and has been alike attentive to the comfort and well-being of the meanest insect, as well as to that of the highest of His creatures. In fact, if we examine through a microscope the smallest insect, we shall find them possessed of the power of seeing, smelling, tasting (for they have a tongue), and hearing. This last faculty is easily ascertained by tapping on the outside of a hive of bees.

After what has been said, I think you will be disposed to own that in no part of His works has the All-wise Creator more vividly displayed His goodness as well as His power than in these atoms, if I may call them so, of creation. They are equally worthy of the study of the Christian and the naturalist, as well as affording instruction and entertainment to those I see around me.

I will now give you an instance in proof of this. You must all of you know the common egg-urchin, or sea-egg, or, as I believe it is sometimes called, the "Sca'ad man's head." Its structure is most wonderful. It is provided with tubercles,—one large and three or four small ones. One small tubercle will separate in rows of pairs—three pairs in each row. Small ridges and furrows also separate the pairs of rows from each other. From the pores protrude suckers, which are very long. The number of these suckers is very great. In a moderate-sized urchin you may count sixty-two rows of pores in each furrow or avenue. Now, as there are three pairs of pores in each row, their number, multiplied by six, and again by ten, would give the great number of 3720 pores. The structure of the sea-urchin is not less complicated in other parts. There are above three hundred of what may be called plates of one kind, and nearly as many of another, all dove-tailing together with the greatest nicety and regularity, having on their surfaces above 4000 spines, each spine being perfect in itself, and each having a free movement in its socket. Its shell is composed of 10,000 distinct pieces, so accurately joined that the whole seems a single shell. Surely we may exclaim that the skill of the Great Architect of Nature is not less displayed in the construction of the sea-urchin than in the building up of a world!

Let me now talk to you a little of the star-fishes, or, as they are called in Ireland, " devil's fingers and devil's hands," where they are sometimes collected in great numbers and used as manure in gardens,—and an excellent manure they make. Now, I am aware that they are no favourites with you fishermen. In my younger days I used to like to go out early and catch a whiting or two for my breakfast. Suppose me seated in a boat, and, after a pleasant pull of half-a-mile out to sea by a good honest fisherman, then preparing for sport. The water seems to laugh and sing, as the Psalmist described the waving corn, and sparkles in the morning sun, and dances around as the anchor—a heavy stone—is cast amidst the gently-swelling waves. Eager for the sport, the lines are prepared; there is a tug at the hook, and, hauling up the lengthened string, expect a prize; when lo! 'tis but a villainous star-fish that has seized the bait, gorging it deep and fast. Again we try, and with the same result. Bait after bait is thus devoured, till I begin to think that shoals of star-fishes are waiting there on purpose to annoy me. No whiting for breakfast! Who would keep his temper under such a trial?

But still they are some of God's useful creatures, and that to a great extent. The appetite of the star-fish is for carrion. Their restless industry is constantly employed in hunting out and swallowing all dead and tainted matters that approach the shore, and which, if permitted to accumulate, would soon pollute the very ocean, defile the air, and thus render the earth almost uninhabitable. Silently and quietly the work is done by these unwearying agents. Cleanliness and health attend their operations, while each, like a living granary, pours forth innumerable eggs, and thus supplies abundant food for countless hungry mouths, peopling the sea with life, and thus adding to the boundless store of nature's provisions.

I have said that fishermen have a great aversion to all sorts of star-fish, or at least it was so formerly, and perhaps is so at the present time. Now these creatures have a wonderful power of reproduction. If one of its arms, or, as they are called, rays, and sometimes fingers, should be taken off, another would come in its place; and in this way, if it is pulled in several pieces, each piece will in time become a perfect star-fish. This may surprise you, but it is a fact well known to naturalists. Thus fishermen were formerly in the habit of tearing them asunder and committing their lacerated bodies to the waves. This plan was anything but the way of diminishing their numbers. You have probably now become wiser, and are content with throwing them on the beach and suffering them to die there.

Another interesting marine animal is the hermit crab, with which you are all well acquainted. In order to comprehend the shape of this creature it is necessary to deprive it of its habitation. It will then be seen to be formed of two distinct portions, the head and a good part of the body being covered with shell, like the fore part of a lobster, whilst the hinder or tail portion is bare, soft, and without any solid protection; so that, in order to defend its hinder regions, the creature is obliged to have recourse to the strange expedient of procuring a retreat in any shells that from their size and shape may be adapted to such a purpose, and which the occupants drag about with them on all occasions. I know that it is a question amongst you fishermen whether the crab selects an empty shell for his purpose, or ejects the lawful tenant of one which he takes a fancy to, by seizing his victim—the whelk, for instance—behind the head, and after killing it proceeding to eat it out of house and home, and then taking possession of the vacant residence.

The form of this crab is wonderfully adapted to its mode of life in a shell. It will be observed, that of the two claws with which it is furnished, one is exceedingly small when compared with the other, and this is a wise arrangement of the Great Creator. Had the two claws been of equal size, both of them as large as the biggest of them, it is evident that they could never have been drawn into the shell. When alarmed, they draw in the smaller claw and close the opening with the larger one, which is thus protected. When the hermit crab grows too large for its shell, it may be observed crawling along a line of empty shells left by the last wave. They then slip their tails out of the old house into a new one, and in this way they will try a number of shells till they find one to their liking. They feed on any kind of putrid offal and garbage, and thus become nature's scavengers in cleansing our coasts.

And now let me say a few words to you respecting the prawn, perhaps one of the most exquisitely constructed of all marine creatures. They are, however, not to be judged of as you may see them boiled and dead in Mr. Hayllar's well-frequented shop; but in a glass-case called an aquarium, if they are well supplied with proper food and fresh sea-water, they will appear the merriest and happiest creatures possible. But it is to the way in which they get rid, once a-year, of their old external covering, that I wish to call your attention. When this period arrives, the prawn ceases to feed, and goes about from place to place until it has fixed on one adapted for its purpose. It then stretches out wide apart its third, fourth, and fifth pairs of legs, and the feet are then hooked on so firmly upon a substance near, and in such a way that the body may be poised, and capable of moving freely in all directions. The prawn then slowly sways itself to and fro, and from side to side, apparently for the purpose of loosening the whole surface of the body from the skin or covering; the two pairs of legs are at the same time kept raised from the ground, stretched forwards, and frequently passed over each other with a rubbing motion. The eyes also may be observed to be moved within their covering, from side to side, by muscular contraction; and when every precaution appears to have been taken for the withdrawal of its body from the old skin, a crack is observed to take place between it and the abdomen, at the upper and back part, and then the head, antennæ, or, as they may be called, feelers, legs, feet, and all their appendages, are slowly and carefully drawn backwards, and out from the dorsal or back covering, until the eyes are quite clear of the body shell, and appear above its margin. The prawn, thus half released, then makes a sudden backward spring or jerk, and the whole of the skin is left behind, generally adhering by the shell of its six feet to the substance it had selected for its purpose.

Now, one moment's thought will show you what a truly wonderful process this act of getting rid of its old covering really is. When we reflect on the small size of this creature, and the extreme delicacy of its various organs, and then find that this moulting of the shell, with its minute spines and microscopic hairs, is performed in the manner you have heard, it is impossible not to admire the wonderful power of an Almighty Creator who has called into existence so marvellous a creature.

When the prawn has been thus liberated from its old covering, it is at first perfectly helpless, and so soft that it has not the power of supporting its own weight. By degrees it gains strength, and then retires to some secure place till its different membranes have become sufficiently hardened to allow of its venturing forth among its companions without danger.

But it is time to conclude. You will observe, that one of the objects in my lectures is to draw your attention to the goodness and wisdom of Almighty God, as they may be seen in His works. These are so various, so beautiful—so perfectly adapted to the purposes for which they were created, that the very knowlege of them should make you better men and better Christians. Be quite sure, then, that there is a God Almighty, all wise and all good, and if we do not shut our eyes we may see Him in all His works, and learn not only to fear Him for His power, but to love Him for the care which He takes of us and of all His created beings.

IX.

ON REPTILES.

My dear Friends,—

I have not, in any of my lectures, said anything on the subject of reptiles, such as turtle, alligators, vipers, snakes, frogs, &c.; and yet I may, perhaps, amuse you by entering a little into some of their history, for many interesting facts may be related of them. These reptiles have their use in the arrangements of Providence, and you should never lose sight of the fact, that nothing has been created but what is for the general good.

Amongst the larger of what may be called reptiles, and which are occasionally found on our coasts, may be mentioned the turtle. The habits of this creature are interesting. At the Isle of

Ascension, and in many other places, innumerable multitudes of turtles arrive in the early part of summer, resorting to their favourite breeding-places. Some come from a great distance. On first nearing the shore, and generally on fine, calm, moonlight nights, the turtle raises her head above the water, when about forty yards from the beach, looks around her, and should she see nothing likely to disturb her, she sends forth a loud hissing sound, and then advances slowly towards the beach, crawls over it, and, when she has reached a place fitted for her purpose, looks all around her in silence. She then proceeds to form a hole in the sand, which she does by removing it from under her body with her hind flappers. The sand is thus raised alternately by each flapper until it is heaped up behind her. In this way a hole is dug to the depth of from a foot and a half to two feet, and which is done in about nine minutes. The eggs are then dropped into it, one by one, in regular layers, to the number of a hundred and fifty to two hundred. This is done in about twenty minutes. She then scrapes the loose sand back over the eggs, and so levels and smooths the surface that few persons on seeing the spot could suppose that anything had been done to it. After this operation she quickly returns to the sea, the eggs being hatched by the heat of the sand. Each turtle has generally three layings of eggs in the season. When the young ones are hatched, which takes place from a fortnight to three weeks after the eggs have been deposited, they make their way to the water, when numbers of them fall a prey to birds, or are seized in the sea and devoured by shoals of fish and crocodiles. However, as we have seen that the female turtle deposits her eggs—probably altogether to the number of five hundred—three times a-year, we may suppose that many escape from their enemies, and thus a provision is made for keeping up a due supply of these useful creatures. This is very much the case with respect to the guinea-fowl. In the extensive woods of Africa these birds are found in great numbers. They lay their eggs on the ground to the amount of twenty or thirty, and the shells of these eggs are so hard that it is not easy to break them. Now, as snakes, which will feed on eggs, are very numerous in these woods, it is evident that, except for the number of eggs laid and their hardness, they would be devoured by the snakes, who occasionally remove some from the nest in trying to break them. I have always looked upon this fact as a beautiful arrangement of Providence for the preservation of His creatures.

I will conclude my notice of the turtle with the following anecdote, which was told me by the late Lord Adolphus Fitzclarence:—In the Island of Ascension turtles are kept in considerable numbers in tanks, to supply the ships which call there. Lord Adolphus commanded a frigate which, on its way to England, touched at Ascension and shipped a number of turtles; but as the voyage home was protracted, many of them died, and were heaved overboard. One large one, called Lord Nelson from its having lost a flapper, survived till the ship was in the Channel, when it appeared so nearly dead that it was thrown into the sea. All these turtles were marked on the shell, as usual, with a hot iron. In the course of a year this turtle was again taken in the Island of Ascension, and was immediately known by the marks as one of those sold to Lord Adolphus Fitzclarence. It is a curious fact, for the animal must have made its way through some hundred miles of sea.

The largest reptiles are the crocodile and the alligator; and, luckily for sea-bathers, not known in this country. Their history is, however, curious, and their habits worth noticing. I will begin with the crocodile. This reptile sometimes attains the length of twenty-five feet, and is by many supposed to be the leviathan of Job, as mentioned in the 41st chapter, and also in the Psalms. It swims rapidly, is very dangerous, and constantly seizes and feeds on human beings. When the British had a detachment of soldiers and some artillery on the banks of the mouth of the river Indus, in the East Indies, a large old crocodile carried off two or three natives, one of them being a woman. Its skin was so thick that no ball penetrated it, so some young artillery officers formed the following plan for destroying it:—They killed a sheep, and in its body placed a bag filled with gunpowder and some other combustible matter, to which a long wire was attached, with detonating powder at the end. Presently the crocodile saw the prey and seized it, and carried it to a hole which he

I

was known to frequent. Time was allowed him to swallow the sheep, when the wire was pulled—the water then became violently agitated—a loud report was heard, and up came the crocodile dead, and his stomach blown open. It is a curious fact, that in the Nile no crocodiles are found in certain degrees of latitude, but they are between 26° and 28°. Cairo is 30°, where they are never seen.

The female deposits her eggs in the sand, about a hundred in number, and nearly the size of those of a goose. An animal called the ichneumon has long been famous in Egypt, where it goes by the name of Pharaoh's rat. It hunts for, digs up, and devours the eggs of the crocodile, thus preventing too great an increase of these dangerous reptiles.

Mr. Curzon, in his travels to visit the monasteries of Egypt, gives the following pleasing and interesting account of his adventure with a crocodile. He says:—

" I had always a strong liking for crocodile shooting, and had killed several of them. On one occasion I saw, a long way off, a large one, twelve or fifteen feet long, lying asleep under a perpendicular bank, about ten feet high, on the margin of the river. I stopped the boat at some distance, and noting the place as well as I could, I took a circuit inland, and came down cautiously to the top of the bank, whence with a heavy rifle I made sure of my ugly game. I then peeped over the bank, and there he was, within ten feet of the sight of the rifle. I was on the point of firing at his eye, when I observed he was attended by a bird called a ziczac. It is a species of plover, and as large as a small pigeon. The bird was walking up and down close to the crocodile's nose. I suppose I moved, for suddenly it saw me, and instead of flying away, jumped up about a foot from the ground, screamed " Ziczac! Ziczac!" with all the powers of his voice, and dashed himself against the crocodile's face two or three times. The great beast started up, and immediately spying his danger, made a jump up, and dashing into the water with a splash which covered me with mud, he dived into the river and disappeared. The bird, proud apparently of having saved his friend, remained walking up and down uttering his cry with an exulting voice, and standing every now and then on the tips of his toes in a conceited manner." The circumstance of the crocodile being often attended by the ziczac has been doubted by some naturalists, but the above anecdote would serve to prove the truth of it. Indeed, Herodotus, an ancient historian, who is generally to be depended on, states that all beasts and birds avoid the crocodile except a small bird which he calls the trochilus, with which the crocodile is always at peace, for he receives benefit from it. When the reptile gets out of the water on land he opens his jaws, when the trochilus enters his mouth and swallows the leeches which infest it. The crocodile is so well pleased with this service that it never hurts the bird. A celebrated French naturalist, after investigating the subject, thought that there was good foundation for the story of this ancient writer.

You may ask, Of what use are crocodiles? They play their part, and that an important one, in the economy of nature. They are to the great rivers of the tropics what wolves and hyenas are to the land and the sharks to the sea. In fact, scavengers clearing away offal and carrion, which would poison the waters and taint the air.

But it is time to turn to the alligator, a variety of the crocodile, although differing from it in many respects. In the first place they make an incredibly loud and terrifying roar, especially in the spring, their breeding season. In the great river Amazon, where these creatures abound, when hundreds are roaring at the same time, it resembles thunder. Unlike the crocodile also, the female makes a nest in the shape of a cone, four feet high, and four or five feet in diameter, constructed of alternate layers of eggs, and of mud, grass, and herbage. When the young are hatched the female tends them as a hen does her chickens. Their cry is like the whining and barking of young puppies.

It would appear, from various accounts, that alligators are much more ferocious than crocodiles; in fact, a much more dangerous reptile. For instance, in one of the Manilla Islands a man rode his horse across a river in a place known to be frequented by an enormous alligator. He got half way over the stream when the alligator came upon him. His teeth went into the saddle, which he tore from the horse, while the rider tumbled over the other side into the water and made for the shore. The alligator, disregarding the horse, pursued the man, who safely reached the bank, which he could easily have ascended; but, rendered foolhardy by his escape, he placed himself behind a tree which had fallen partly into the water, and, drawing a heavy knife, leaned over the tree, and, as the alligator approached, struck him on the nose, and kept on repeating the blows until the animal, exasperated at the resistance, rushed on the man, and seizing him by the middle of the body, which he crushed, swam into the river, where the poor man's sufferings could not have been very long. This alligator was soon afterwards killed by means of being drawn into three very strong nets placed in the river: he broke through two of them, but got entangled in the third, and was speared to death after a long resistance. This tremendous brute was nearly thirty feet in length and sixteen feet in circumference, and his head alone weighed three hundred pounds.

A young girl, about thirteen years of age, was washing a towel in a river frequented by alligators. She did not attend to a warning to beware of them, and just as she was boasting that she did not care for them, a scream for help was heard, and a cry, " Lord, have mercy on me!

—alligator has caught me!" The body was found, some days afterwards, half devoured.

An English naturalist, who was in search of plants in South America, says, " I was disappointed not to observe a single plant, except the rank grasses, round the margin of the river; but alligators were laid in the water in almost countless numbers, resembling so many black stones or logs. What we had seen in the river Amazon of these reptiles was nothing compared to their abundance in the Ramos river and its adjoining lakes. I can safely say that at no one instant, during the whole thirty days, when there was light enough to distinguish them, were we without one or more alligators in sight." In the lakes, towards the close of the rainy season, myriads of ducks breed in the rushes, and here the alligators swarm to feed on the young birds. If a sportsman fire at the ducks in these places, he has but a poor chance of bagging many; for the instant a bird falls on the water the alligators rush towards it and crash their huge jaws upon each other's heads in their hasty attempts to seize it. When alligators have been hungry, they have been known to upset a small boat in order to feed on the rowers. Many instances have been recorded of persons having lost an arm or a leg, which has been taken off by these monsters.

The alligator of North America buries itself at the bottom of marshes till the spring sets in, and it is then in such a state of torpor that slices may be cut from the animal without arousing it. On the other hand, the alligator revels in the moist heat of Florida, and is formidable, both in numbers and size, at a mineral spring near the Musquito River, where the water, on issuing from the earth, is not only nearly boiling, but is strongly impregnated with copper and vitriol.

I will now talk to you about that poor, persecuted, but harmless and useful reptile, the toad, and I hope to rescue it from some part of that ill-treatment it so constantly meets with. It is a timid creature, perfectly inoffensive, and, as I know, will attach itself to those who show it kindness. Mr. Bell tells us that he had a large one that would sit on one of his hands and eat from the other. A true lover of Nature, in that simplicity and singleness of heart which always belong to that character, will find in the toad much to admire, although it must be confessed it has an ugly appearance. It is of great use in a garden, devouring great numbers of slugs, worms, and destructive insects. It is kept in cucumber and melon frames for that purpose. I got a friend to send me twelve toads in a box from Jersey for the late Mr. Knight, of the Exotic Nursery, Chelsea, for one of his stove-houses, which was much infested with insects, and they cleared it of them very quickly. It darts its long tongue out when two inches from its prey, and seizes it so rapidly that it is difficult to perceive its motion. Three gentlemen, while walking on the Fairlight Downs, near Hastings, saw a toad squatted on the ground, holding the head of a viper in his mouth. The viper writhed its body as if trying to escape, but to no purpose. The eyes of the toad glared, and it showed much ferocity. The entire head of the viper was in the toad's mouth, which seemed completely filled in consequence, and its jaws were closed, and yet it appeared to breathe freely. This is the only instance I have ever heard of a toad attacking a viper.

The toad, like the snake and other reptiles, sheds its skin at certain periods; but not until a new one has been formed underneath. The old skin then cracks along the back and belly, and after a few struggles and shakes the toad is free.

It is a curious fact that, although toads abound in the Island of Jersey, they are never found in the neighbouring island of Guernsey, and, if imported into it, always die.

Much has been said and written of live toads being found in blocks of stone and in trees; but I have not been able to procure an authenticated fact of this circumstance: nor is it likely that they would live for hundreds of years in such situations as has been confidently stated. I once put a toad in a flower-pot, which I placed on a flat tile, stopping up the hole at the bottom of the pot, and buried them about a foot deep in the earth. At the end of the year I released the toad, and found him as well and as lively as before his imprisonment; but this is no argument of their living many years in a block of stone. It is torpid in winter, and then retreats into some sheltered spot, and there remains till the return of spring.

The frog is, perhaps, a more interesting reptile than the toad. It is a harmless as well as a very useful creature, serving Frenchmen for food, but living itself on various insects and slugs, which it devours in large quantities, so that it should always be encouraged in gardens. In winter, they congregate in multitudes, generally in the mud at the bottom of the water, adhering together so closely that they appear like one mass. They separate on the return of spring; then their cheerful croak is again heard, and they recommence their active life.

Frogs have been supposed, as far back as the times of good old Izaac Walton, to have a great antipathy to pikes, killing that fish whenever it can. There is some truth in this. A gentleman, walking in the spring on the banks of a piece of water at Wimpole, the seat of Lord Hardwicke, observed a large pike swimming in a very sluggish manner, near the surface of the water, having two dark-coloured patches on the side. A few days afterwards he saw the same pike floating dead upon the surface of the water, and, having drawn it to land by means of a stick, he found that the two dark-coloured spots were two living frogs, still attached to the fish, and that so firmly that it required some force to push them off with a stick. These reptiles are so well known that little more need be said of them. They are a favourite food of the common snake; and while the snake is endeavouring to swallow a frog, the cry of the latter is very loud and distressing.

But I must conclude. I have done my best to amuse you, by setting before you some of the Great Creator's works, all of which have their peculiar uses assigned to them. Let us also be thankful that this happy country is not infested with those reptiles which you have heard abound so much in hotter countries. This is one, among many blessings, bestowed on this land.

X.

ON THE HABITS OF ANIMALS.

My dear Friends,—

I am going to read you a lecture on the habits of animals generally, and hope it will amuse you. One of my objects is to do so, and another to instruct you. In fact, the instincts of animals, their contrivances, their architecture, their forethought, their affections, and various other circumstances connected with their several modes of life, are, indeed, lessons of instruction to every one. They show the goodness of the Great Creator. They serve to prove the truth of what the Psalmist said, "The eyes of all wait on Thee, and Thou givest them their meat in due season." If the most insignificant little living creature is viewed through a microscope, it will be found to be most exquisitely formed, and to possess all those functions which are necessary to its well-being. I wish to impress this strongly on your minds, that you may be able to view Almighty God in His works, and thus learn to love and admire Him as the Author of all good. Having stated this, I will now proceed to give you some curious instances of instinct amongst what are called the inferior animals, that is, of animals which are supposed to be almost without sense or power of motion.

You are all of you acquainted with the common sea-hog, or sea-egg. To look at it you would suppose it to be without sense, or the possibility of regarding external objects by sight or hearing; yet it will travel up the rods of a crab-pot, enter the opening, descend within, mount again to the situation of the bait, and choose the one which pleases it best.

Again, the star-fishes seem very inactive, and without intelligence; yet they display sagacity in the discovery and choice of food, as well as in the manner of seeking it, and also alter their habits in different seasons.

You might suppose that cuttle-fishes were

without any sense, and yet they show some degree of curiosity by their moving up to a shining object to examine it; and when in danger, they become suddenly suffused with a decided blush of red, and then eject the contents of their ink-bag, by which they become concealed from observation.

The oyster closes its shell when it comes in contact with some objects, and opens it on the flowing of the tide, so that its structure is wonderfully adapted to the wants and circumstances of a creature so apparently unconscious of a want, or, if it had one, so incapable of supplying it.

But let me turn to a higher order of animals; and I will begin with birds, and their instinct of migration. You are aware that numbers of birds arrive in this country in the spring from far-distant regions, flying over immense tracts of land and sea in one unerring line, and generally doing this in the night. This instinct induces them to seek a warmer climate at one season of the year, and a colder at another. It cannot be supposed that the old birds lead the young ones in these migrations, for it has been ascertained that late broods have taken their flight long after their parents have departed. Indeed, the young cuckoo has never known a parent's care, since it is brought up in the nest of some other bird; yet it leaves this country long after others of the same species, being then a solitary individual, and finds its way to the groves of Greece and the sunny regions of Italy. Now, it is quite clear that this extraordinary migratory instinct must have been implanted in this and various other birds by a merciful Creator, for purposes intended for their well-being in climates congenial to their respective wants. And then, with what pleasure may we listen to the songs of numerous warblers which arrive amongst us in this blessed country, the nightingale being amongst them, cheering us with their music, and proclaiming the loving-kindness of our Heavenly Father!

But this migratory instinct is not confined to birds. The extensive plains of North America were formerly more frequented by vast herds of buffaloes than they are at present, in consequence of the destructive attacks made upon them, not only by the Red Indians but by the American settlers. At certain periods a strong migratory impulse seizes upon those great herds, and they rush along the plains, cross rivers, ascend hills, and go in one undeviating line to some far-distant locality, overturning tents and other obstructions in their way. So it is, also, with the land-crabs of Jamaica. When the season arrives they quit the upper country and make a rush towards the sea-shore in a direct line, and nothing stops their progress, so strong is the migratory instinct.

This extraordinary impulse is also possessed by some insects. In Australia a migratory procession of caterpillars may frequently be observed. They travel in single file, having a leader; and each is so close to its predecessor as to convey the idea that they were united together, moving, like a living cord, in a continuous line. If one caterpillar should be taken from the middle of the line, the one immediately before him suddenly stands still, then the next, and then the next, and so on to the leader. The same takes place at the other end of the line. When the caterpillar which is removed gets into the line again, the whole move forward as at first, thus apparently having some means of communication with the leader.

This was not a solitary experiment, but has been repeatedly tried during the progress of these insects, and proves the extraordinary fact of the power of communication existing amongst them.

Quails have a strong migratory instinct, and so regular is their arrival in the island of Malta, that the day of their coming is noticed in the published calendar of the island, as the change of the moon is in ours. A great flight of storks also takes place annually in the Mediterranean, about the same period of the year; and I was assured by the captain of a ship, who was engaged in making surveys on the coast, that those young birds which were incapable of performing so long a flight during the migratory impulse, were conveyed on the backs of their parents to far-distant places, some of them making their way into Persia.

One of the most curious instances of migration is in the case of the heron. Heronries are not very common in England, and certainly there are not any within some miles of Richmond Park, in Surrey. Yet year after year (for I reside in the neighbourhood) I have seen from fifty to sixty herons assembled on a large open space in that park, not moving about or seeking for food, but appearing as if they had met together to consult on some important subject. What their object was I never could even guess at; and it must be some extraordinary instinct which brings them thus together, especially as no heronry has so many of these birds belonging to it, so that the assembly must consist of several heronries.

It is an interesting fact, that instinct leads migratory animals in general to pursue one invariable direction in their passage from one distant country to another. They have neither compass nor guide, and yet they rarely deviate to any great extent in their journey. Mountains and wide seas intervene, and yet young and old find their way. Inscrutable as this instinct may appear to our dull perceptions, it is implanted in His creatures by a wise and good God, who leads even the little, feeble insect-hunting birds to go remote distances from their homes to seek for that food which they require. Indeed, their rapid flight proves that they have a conscious security of finding it. A lady told me, that while cruising in her husband's yacht in the wide seas of South America, she witnessed a migration of butterflies, which were far distant from any land, and on a subsequent occasion a migration of humming-birds. Change of weather, no doubt, produces migratory impulses—a fact which some of you fishermen are acquainted with.

Some few years ago immense cloud-like swarms of dragon-flies passed in rapid succession over a town in Germany. Their progress was from south-by-west to north-by-east, some flying high and others low, and they struck against the windows of houses situated on eminences. We are not visited in this country by locusts, which commit such vast injuries on crops in the East. When the migratory impulse is on them they swarm in vast numbers, taking long flights, and sometimes alighting in the sea and perish.

Some animals change their quarters (it may be called migration) for unaccountable reasons. For instance, the badger, which is a solitary animal, and once very numerous in this country, would assemble to the number of nine or ten, and travel by night to some other locality. If any one happened to disturb them in their progress, he was attacked with the greatest ferocity, although when single they are perfectly harmless. Rats and mice are also known to migrate; but always by night. Speaking of rats, I may tell you that a gentleman in Herefordshire laid up about two bushels of walnuts, and on the following morning he found that they had all been carried away by rats.

But I should tire you if I were to continue longer the subject of instinct in animals, although I might pursue it to a great extent. In many cases it amounts so nearly to reason, which latter faculty is supposed to be possessed only by the human race, that it is difficult to define where instinct begins and reason ends. You may judge for yourselves when I give you the following anecdotes:—

A fox, partly tamed, was kept fastened by a chain to a post in a court-yard, and was chiefly fed on boiled potatoes. Many fowls also were kept in the same yard, but had sense enough not to come within reach of the fox. He was, however, too cunning for them, as you will find. Having bruised and scattered the boiled potatoes which he had received for his dinner at the extremity of the space the chain would reach, he retired to an opposite direction, and put on the appearance of being asleep. His cunning succeeded, for some of the fowls were thrown off their guard, and came within the circle of danger to eat the potatoes. The fox then sprang upon them and seized his prey. Was there not some degree of reason in this?

Again: an old man was walking one day upon the banks of a river, when he observed a badger moving leisurely along the ledge of a rock on the opposite bank. In a little time a fox came up, and after walking some distance close in the rear of the poor badger, he leaped into the water. Immediately afterwards came a pack of hounds in pursuit of the fox, who by this time was far enough off floating down the stream, but the unfortunate badger was instantly torn to pieces by the dogs. Here was cunning combined with reason. A fox has been seen to drop the end of his tail among rocks on the seashore in order to catch the crabs below, hauling up and devouring such as laid hold of it.

I will now give you an instance of what might be called reason in a dog, and which occurred in this town. A lady, proceeding to the house of one of her pupils, near Brunswick Place, had her cloak seized by a dog, that pulled her the contrary way to which she was going. As she could not disengage herself, she permitted herself to be led till she was brought to the open space at Wick, when she became alarmed, and asked some men to drive the dog away. They persuaded her to see where the animal would lead her, promising to protect her if necessary. He brought her to a house which was then in the course of erection, and began to scratch at the end of a plank, which was laid across the open unfinished area of the house for the workmen to get into it. The plank was lifted up, and a beef bone found under it, which the dog seized and ran away with. This dog belonged to an excellent, charitable clergyman at Wick, who told me the anecdote after he had taken some pains to ascertain its accuracy, so that it need not be doubted. It is a curious instance of a reasoning faculty in an animal. But I must now conclude.

You will recollect that I have given you several lectures on Natural History, and I wish you to consider them, not as a mere gratification of curiosity, or as vehicles for amusing anecdote, but as affording proofs of a Superintending Providence, and of the care bestowed on all the works of creation by a Being, infinite in power, wisdom, and goodness! Indeed, he must be wilfully blind who does not observe Divine interposition, not only in human affairs, but in everything connected with the animal and vegetable creation. With respect to the vegetable creation I may tell you, that if potatoes are put in a cellar and the least ray of light admitted, the shoots of that vegetable will move in a direct line to that light. Again: if the root of a plant be uncovered, without exposing it to too much heat, and a wet sponge is placed near it, but in the opposite direction from that in which the root is proceeding, you will see the root turn towards the sponge as if wanting to imbibe its moisture.

Lastly, place an upright pole near an unsupported vine which is growing in an opposite direction to the pole, it will quickly alter its course, and stop not until it has fixed itself to the pole. These facts may excite our astonishment, but they should, at the same time, produce in us feelings of gratitude, and, I may add, of confidence towards that benign Being who supports us as well as the humble vine, and showers down His blessings upon us.

XI.

THE GOODNESS OF GOD.

My dear Friends,—

I am glad to meet you again, and to see those faces around me which I have so frequently looked upon with pleasure. That pleasure is increased by the feeling that I may not only have amused you, but possibly have done you some little good, inducing you to feel that, although your occupation leads you on the great waters, you have an immortal soul to attend to. You fishermen ought, indeed, to be more conscious than landsmen that you are in the Divine Presence, and have been constantly indebted to Divine protection. When at sea, you have little to look upon but the heavens above or the boundless ocean around you. Both may seem to be created on purpose for you—the sun to guide you by day, and the stars by night; the sea to bear your boats on its bosom, and the breeze to waft you on your course. You probably feel how powerless you are of yourselves—how frail your vessels—how dependent you are on the goodness and mercy of your Creator, and that it is He alone who can rule the tempest and control the stormy deep. The ocean for awhile separates you from the vices and temptations of the world, and enables you to reflect that God is the author of all you see around you.

I have entered into these remarks, because I wish in this lecture to give you some proofs of the goodness of the great Creator in furnishing every animal with those habits and that clothing proper for the country in which they live, and also how admirably the structure of their bodies is adapted to their peculiar way of life. Thus, for instance, elephants, rhinoceroses, and monkeys feed upon vegetables and fruits that grow in hot countries; and such places are, therefore, allotted to them. No sun is too powerful to hurt them, and they do not need hair or wool to keep them warm. On the contrary, the reindeer are found in the coldest part of Lapland, and they are covered with the thickest hair, and thus can defy the severity of the winter. In like manner, the rough-legged partridge passes its life in the Lapland Alps, feeding upon the dwarf birch; and that they may be able to run about safely amidst the snow, their feet are feathered.

The camel frequents the sandy and burning deserts, nor could they be passed without him; but how wisely has the Creator contrived for him! In traversing the deserts, where no water is to be found for many miles, and where every other animal would die of thirst in such a journey, the camel can undergo it without suffering, for his stomach is full of cells, in which he reserves water for many days.

The pelican also lives in deserts and dry places, and frequently builds her nest far from any water, in order that her eggs may be hatched by the heat. She is, therefore, obliged to bring water from afar for herself and her young, for which reason Providence has furnished her with a very large bag under her throat, which she fills with a quantity of water sufficient for many days.

The feet of goats are admirably adapted to

enable them to climb over rocks and the precipices of mountains.

Swine, especially in their wild state, have very strong powers of smelling. Thus they are able to find succulent roots in the ground, which they turn up and feed upon.

Squirrels are so formed that they can climb up trees with great rapidity, and so are woodpeckers.

Swallows are beautifully made for their peculiar mode of life. Thus the shape and lightness of their bodies, and the length of their wings, not only enable them to fly from morning to night, in search of flies and other insects, but also to take long flights across wide seas and distant lands to different climates, according to the seasons of the year.

Look, again, at the mole, how curiously it is made for its underground life! Its fur is close, thick, and soft; its feet are admirably made to enable it to form its runs in the earth; its eyes are so small that nothing can injure them; and its powers of smelling so acute, that it can detect worms and other insects under the soil on which it feeds. Its sense of hearing is so acute, that if a footstep approach the spot where it is at work, it will immediately cease its operations. It is a most useful animal, although many farmers are so foolish as not to have found it out.

But it would be an endless employment if I were to enter fully into the various ways in which a benevolent Providence has provided for the wants of His creatures by their peculiar formation. I will now endeavour to give some instances of those peculiar instincts which some animals possess, and which tend to their self-preservation, or that of their young. For instance, when a female otter has been attacked in company with her young one, she will clasp it with her forefeet and plunge beneath the surface. Instinct tells her, that although she can remain for some time under water herself, her young one cannot, and, therefore, she is forced to rise again very soon. Her love for it is so strong, that if it is taken its cries bring her to the side of the boat, where she often shares the fate of her cub.

When rooks are feeding, they always place a sentinel on a tree to give an alarm in case of danger. Fieldfares, and other birds which collect together in large flocks, do the same. Instinct alone could have taught them the necessity of doing this.

You have probably heard of the monks of St. Bernard, whose monastery is situated amongst the snow and ice of the Alps. They have a fine, strong, and intelligent breed of dogs, which have been taught to wander over the snow, and seek for travellers who have been buried in it; and thus many lives have been preserved. A friend of mine procured a very young puppy of this breed from the good monks, and brought it to England in the winter. There was snow on the ground when it was turned loose, but, young as it was, it immediately began to scratch in the snow, as if seeking for some one, so strong was its instinct.

We have about forty little, tender, migratory birds, which arrive in the spring, and many of them cheer us with their songs. Now, these birds must fly over wide seas and lands before they can reach our shores; and this is generally done in the night. You may ask, What leads them to do this—to encounter so much danger and fatigue? It is an instinct which Almighty God has implanted in them; and so unerring, it is supposed, must be their flight, that they come in one direct line to this country in the proper season.

When a salmon has been hooked or taken in a net, it immediately discharges the contents of the stomach, as instinct teaches it that it has a better chance of escape with an empty than a full one.

Some animals will put on the semblance of death when their lives are in danger. For instance, the opossum does it; and the common snake will do this, and also a bird called the landrail, and some beetles.

Instinct has taught the peacock, when danger threatens it, to expand its beautiful tail, to shake its quills, and to hiss like a serpent, the shape and colour of its head resembling that reptile; and it must be a bold animal which would attack it in this position.

When a cat gets into a wood, and tries to conceal herself, that she may find an opportunity of seizing her prey, you may hear the screams of blackbirds, and the alarmed cry of numerous other birds, all directing their attention to the spot where the cat is concealed, thus informing the whole neighbourhood of the presence of an enemy. This is the instinct of vigilance.

Instinct shows itself in a variety of ways. For instance, an otter produced a pair of young ones in the Zoological Gardens in London, and these young ones got into a pond when but half filled with water, and were unable to climb up its perpendicular sides. When they had remained in the water some minutes, the mother appeared anxious to get them out, and made several attempts to reach them from the side of the pond. She then plunged into the water, and, after playing with one of them for a short time she put her head close to its ear, as if to make it understand her intention, and then sprung out of the pond, while the young one clung tightly by its teeth to the fur at the root of her tail. Having landed it she rescued the other in the same manner. This offers a curious instance of a communication of ideas between a parent and its young, in consequence of some peculiar instinct; but it also shows the early age in which intelligence is possessed by offspring.

The love of life is possessed in common both by man and animal. In the latter, instinct teaches it to avoid danger, and in man reason is brought into action for his self-defence. A moor-hen, whose nest amongst rushes had been frequently destroyed by a sudden rising of the water in a neighbouring pond, built a nest in a spruce-fir tree near it, at a height of twenty feet from the ground, instinct teaching her that there it would be safe. A similar instance occurred in my own neighbourhood in Richmond Park, where there were some Cape geese. These laid their eggs on an island in the middle of the pond, but they were constantly fed upon by water-rats. Finding this to be the case, the geese made their nests in some oak pollard trees near the water, where they laid their eggs and hatched their brood in safety. They then took their young, one by one, in their bills to the pond. I have known a swan, just previous to a sudden rise of a river, add a quantity of materials to her nest, assisted by her mate, so as to raise it above the flood-mark at least two feet, and thus prevented her eggs being chilled. Creative Wisdom could alone have endowed the swan with this extraordinary instinct of foreseeing a flood and guarding against its consequences.

Instinct has taught a curious little animal in Australia, something between a rabbit and a rat, to collect two or three cartloads of sticks, interwoven in such a way as to form one solid mass, and in this the young are brought forth and reared. The object in doing this is to protect the young from being destroyed by the wild dogs of the country. Instinct has also taught some birds, where monkeys and snakes abound, to build their nests at the extremity of slender branches of trees, in which they lay their eggs and rear their young in perfect security.

At Cape Comorin, the most southern part of Hindostan, there is a bird called the baya bird, which suspends a glow-worm to its nest. These birds are very numerous there, and they have hanging nests. At night each of their little habitations is lighted up by a firefly stuck on the top with a bit of clay. The nest consists of two rooms. Sometimes there are three or four flies on them, and their blaze in the little cell dazzles the eyes of the bats, which often kill and feed on the young of these birds. I have the authority of Dr. Buchanan for this interesting anecdote.

In relating these facts to you, I wish to impress on your minds that no creature, however weak and ignoble it may appear, is left unprovided and defenceless to take its chance in the struggle for existence. Each is endowed by its Creator with bodily and mental attributes, most perfectly adapted to its sphere of action. The humble worm is furnished with innumerable hooks to enable it to penetrate the soil, and to turn up those little hillocks or casts which you must have seen, and which enrich the earth. The spider spreads its beautiful network to catch its prey, showing an industry and perseverance equally extraordinary; while the little bee flits about from morning to night in search of honey, with which to store her hive, hastening from flower to flower: thus fertilising blossoms in her flight, and rendering the fields and gardens gay with flowers and productive of fruit. And here I cannot help quoting part of a speech made lately at Leeds by one of our Ministers, Lord Palmerston, on this interesting subject. He said:—"The contemplation of these organic beings must fill the mind with admiration at the amplitude of the creation, and of the care and skill and wisdom which have directed the Great Creator to whom they owe their origin." It has been my endeavour to prove to you how beautifully and beneficially God's providence acts upon His creatures by endowing them with habits, formation, and instincts necessary for their self-preservation, and which, as you have heard, one of our most enlightened Ministers thought worthy of notice while addressing a large audience of people.

What I have been saying to you is not only for the purpose of proving to you the wisdom and goodness of the Great Creator in every thing we see around us, but to set before you how beautifully and beneficially His providence acts upon His creatures by endowing them with habits, formation, and instincts necessary for their self-preservation. This is a subject well worthy of your consideration; for if God feeds the raven and attends to the cry of her feeble young for food and supplies it, how much more readily will He hear your prayers if they are poured forth with an earnest and humble desire of being benefited by them! I also wish to impress upon you that God's method of government is by rewards and punishments — that is, that He will punish evil and wicked men for their conduct in this world, and reward those, both in this world and also in the next, who with humble and contrite hearts diligently seek Him.

I have one more lecture to give you before I leave this place, and, from my advanced age, it may probably be my last. Whether it may be so or not no one can tell, but this I beseech you, do not forget what your old friend has affectionately and honestly said to you, with an earnest desire of doing you good. Bear constantly in your minds that you are sent into this world for the purpose of seeing whether you will prepare yourselves for a happier and far better one, or live in a state of indifference, wickedness, and hardness of heart, till you die, like the beasts which perish, without hope and without comfort. I pray God that of His infinite goodness and mercy He will preserve and bless you all.

XII.

ON SAVINGS' BANKS.

My dear Friends,—

I am now going to address you on a subject which I think may be of use to you, and I am sure you will give it your serious attention. My wish is to do you good.

You are all of you labourers—that is, you work for your living: whether by sea or land is of no consequence to what I am going to say. Now, I must tell you that I consider a labourer, with the use of his own good right arm, is an independent man. You may be surprised at my saying this; it is true, nevertheless. But, in order to become independent, remember that you must be diligent, thrifty, and sober, avoiding those curses of hard-working men, the ale-house and beer-shop.

But there is another method of becoming independent, and that is by deposits in a savings' bank. Now, you may think it difficult to put by a trifling sum every week out of your earnings; but be assured that you may do this if you have the will. In the poorest family there are odds and ends of income apt to be frittered away in unnecessary expenditure, but which might be saved. I will give you an instance in proof of this. In a poor village in the north of England, a good clergyman established a savings' bank. It was a very unlikely place to succeed; but he did succeed. The wages of the workmen were only 8*s.* a-week, and female labourers and servants had much less. The institution rose in four years as follows:—The first year 151*l.* were deposited; 176*l.* the second year; 241*l.* the third; and the fourth, 922*l.*

Let me give you another instance, also well authenticated. The Royal Artillery Corps has 1432 depositors, and their savings on the 31st of March, 1859, amounted to 23,012*l.*, or an average of 16*l.* to each depositor. This was done out of a daily pay of 1*s.* 3*d.* and one penny for beer money, or about 9*s.* 6*d.* a-week, but subject to deductions for extra clothing, &c.

During the Crimean war the Army Works Corps sent home 35,000*l.* of their savings.

The celebrated George Stephenson, who laid out so many miles of railroad, told me one day that he began life as a poor labouring boy, and that it was a source of great joy to him when his wages were raised to 12*s.* a-week, and he said that he was then a made man. He not only maintained himself upon his 12*s.*, but helped his poor parents, and paid for his own education. When his wages were advanced to 20*s.* a-week, he immediately began, like a thoughtful, intelligent workman, to lay by his surplus money, and when he had saved his first guinea he proudly declared to one of his brother-workmen that he was now a rich man; and he was right. For a man who, after satisfying his own wants, has something to spare, is no longer poor; and a person of great experience has declared that he never knew amongst the labouring class of a man who, having out of his small earnings laid by a pound, had in the end become a pauper. In order to show what diligence and intelligence will do, I may mention that Mr. Stephenson told me some years ago that he had lived to lay out forty millions of money in railroads and other public works, and he died a rich man; but remember, that he began by placing money in a savings' bank.

I will now give you two pleasing instances of the use of savings' banks.

One evening a boy presented himself to draw 1*l*. 10*s*. from the bank. According to its rules a week's notice must be given before any sum exceeding 20*s*. can be withdrawn, and the cashier, therefore, hesitated to make the payment. "Well," said the boy, "the reason's this: mother can't pay her rent; I'm going to pay it, for as long as I have awt she shall hev' it." In another instance a youth drew 20*l*. to buy off his brother, who had enlisted. "Mother frets so," said the lad, "that she'll break her heart if he isn't bought off, and I can't bear that."

You have heard of the London Ragged Schools, and useful and admirable schools they are. Now, in 1859 not less a sum than 8880*l*. was deposited by these poor children, in 25,637 sums by them alone. If this can be done by these children of the ragged schools, how much better are you able to do it! Remember that you have all of you to provide against three contingents: want of employment, sickness, and death. You may escape the two first, but the last must come. It is, however, your duty to provide against the two first of these; and this you may do by deposits in the savings' bank.

If you will not help yourselves, how can you expect others to assist you? But I am sure that a good, steady, provident man, who has earned an excellent character for himself, if he should be attacked by sickness or any other misfortune, is sure to find friends and assistance. Enjoy your pint of beer or hot coffee, but avoid the ale-house—for you there spend, or rather waste, your money and ruin your families; for it has been computed that amongst those who earn a sufficiency to live upon comfortably, one-half of it is too often spent by the man upon objects in which the other members of his family have no share. Now, this is selfishness, to say the least of it.

I cannot conceive a much greater pleasure that a working man, with a wife and family, can experience, than a feeling that he is independent—that is, that he has saved up sufficient money to guard against bad times and sickness. Many of you have good, handsome wives, and pretty children. It is your duty to save up money for them, and you will enjoy your home much more when you have done this. May I ask you a question? When you have had more money about you than you require for current purposes, have not some of you been tempted to spend it? Is it not, to use a common phrase, apt to burn a hole in your pocket? Are you not easily entrapped into company, and into an ale-house, with its bright fire, and there spend your money foolishly, and perhaps become intoxicated? It is a fearful fact, that in 1859 there were, throughout the kingdom, 152,222 houses licensed to sell intoxicating drink, and only 606 savings' banks. Indeed, in Manchester alone, there are 6306 houses licensed to sell drink, and in the large, populous county of Lancaster, only thirty savings' banks. You see, then, what sad inducements there are to tempt the working classes to impoverish themselves. On the contrary, thousands of working men have been benefited by savings' banks, in which large sums have been accumulated. For instance, a respectably-dressed working man, when making a payment one day at the savings' bank, which brought his account up to nearly 80*l*., informed the manager how it was that he had been induced to become a depositor. He had been a drinker, but one day accidentally finding his wife's savings' bank deposit-book, from which he learnt that she had laid by about 20*l.*, he said to himself, " Well, now, if this can be done while I am spending, what might we do if both were saving?" The man gave up his drinking, and became one of the most respectable men of his class. "I owe it all," he said, "to my wife and the savings' bank."

It is my wish to impress upon you that, if you do not lay by money, you cannot improve your present condition. You are fixed like a limpet to the rock. But with some money at your command, you may find various methods of laying it out to advantage. Recollect also, that if you cannot deposit one or two shillings every week, you can put into the Penny Bank, and this generally leads to larger deposits.

I have only to add, that I trust all of you will avoid the great sin of drunkenness. It is a sort of leprosy, clinging to many in this happy country. People may talk of deaths by war, by disease, or famine; but be quite sure that, destructive as these are, they are nothing when compared with the deaths caused by intemperance. People also talk of reforms — such as reforms in Parliament, in religion, &c. — but depend upon it, that the great reform we want is the enfranchisement of our fellow-creatures from the degrading effects of drunkenness. The filth and misery which fester round the drunkard must be seen in order to be known. He is, in fact, worse than a brute to himself, as well as to his wife and children, degrading them and degrading himself. Remember that drunkenness makes some men fools, some men beasts, and some men knaves.

I have now entered my eighty-second year, and may have but few opportunities of addressing you again; but it would be a great happiness to me to think that what I have now affectionately said to you will be borne in mind when I am no more. May God, of His infinite mercy, bless you all.

LONDON:
STRANGEWAYS AND WALDEN, PRINTERS,
28 Castle St. Leicester Sq.

www.ingramcontent.com/pod-product-compliance
Lightning Source LLC
Chambersburg PA
CBHW030244170426
43202CB00009B/622